Liza Pettett

VALENCIA COLLEGE

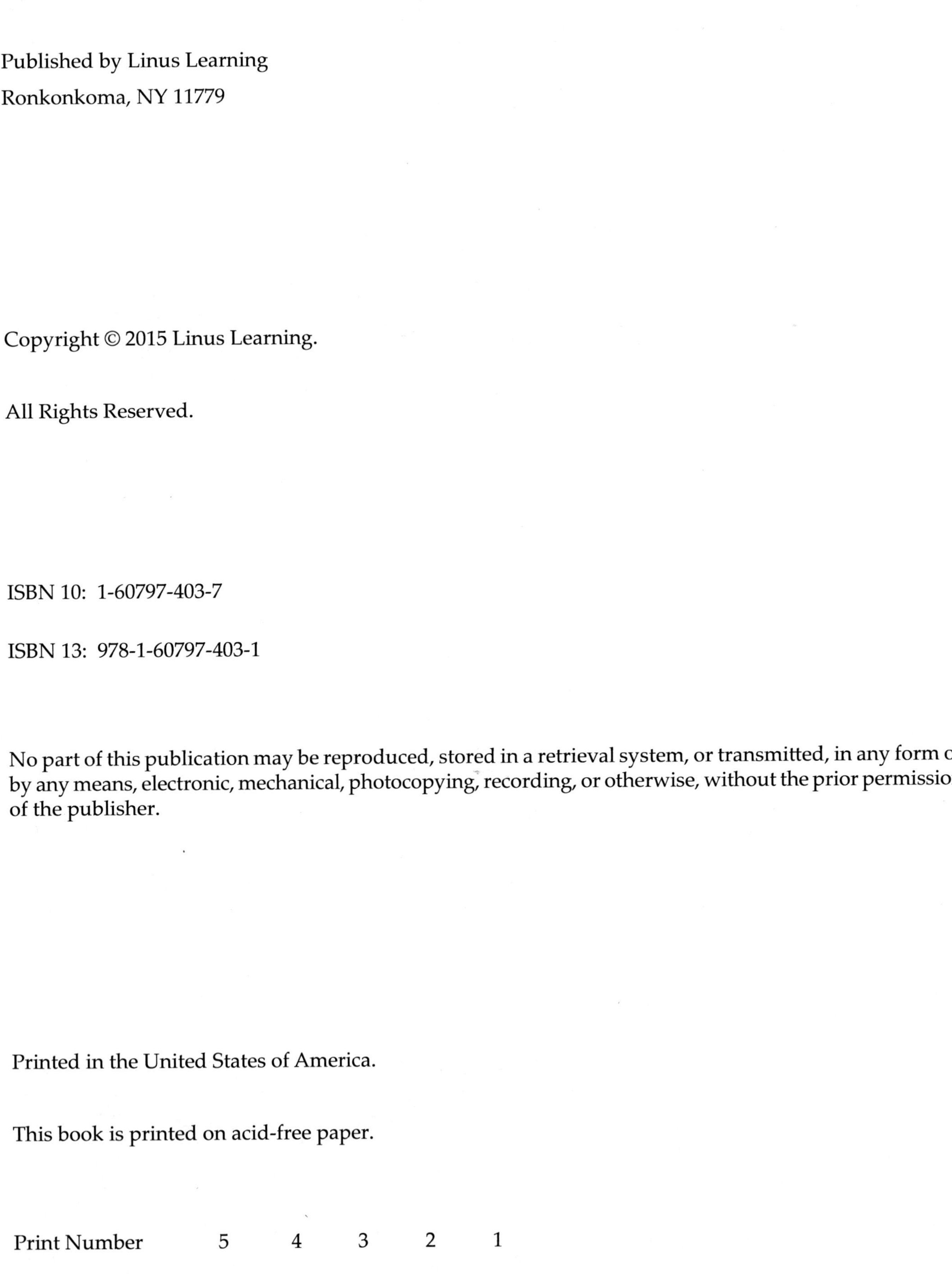

Published by Linus Learning
Ronkonkoma, NY 11779

ISBN 10: 1-60797-403-7

ISBN 13: 978-1-60797-403-1

Printed in the United States of America.

This book is printed on acid-free paper.

Print Number 5 4 3 2 1

TABLE OF CONTENTS

CHAPTER 1: WHOLE NUMBERS

CHAPTER 2: INTEGERS

CHAPTER 3: DECIMALS AND PERCENTS

CHAPTER 4: FRACTIONS

CHAPTER 5: EXPRESSIONS

CHAPTER 6: EQUATIONS

CHAPTER 7: APPLICATIONS

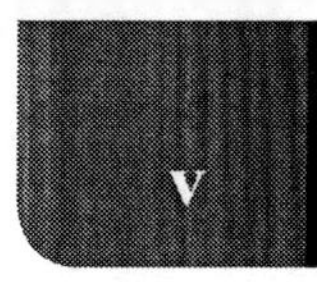

CHAPTER 8: COORDINATE SYSTEM AND POLYNOMIALS

CHAPTER 9: RATIOS AND PROPORTIONS

CHAPTER 10: GEOMETRY

CHAPTER 1

WHOLE NUMBERS

1.1 INTRODUCTION TO THE WHOLE NUMBERS

Objectives:

- Find the Place Value of a digit in a whole number.
- Write a whole number in standard form and word form.
- Write whole numbers in expanded form.
- Read tables and information.

Sets of numbers

A *Set* is a collection of objects. When writing a ***Set*****, braces { }**are used to enclose its **members** so known as elements.

The **set of natural numbers** {1, 2, 3, 4, 5, 6...}; this set does not contain the value of zero.

The **set of whole numbers** {0, 1, 2, 3, 4, 5, 6 ...}; contains the value of zero.

Place value: The position of a digit in a numeral determines its value. When placing in the chart you must start from the last digit and work yourself to the left in the chart.

PLACE VALUE CHART

Billion Family			,	Million Family			,	Thousand Family			,	Ones Family			.	Decimal Family		
Hundred Billions	Ten Billions	Billions		Hundred Millions	Ten Millions	Millions		Hundred Thousands	Ten Thousands	Thousands		Hundreds	Tens	Ones		Tenth	Hundredth	Thousandth

TRY these:

345,576,897,415 and 158,722,000

Expanded notation:

you write the numerals in word form as you would when writing a check.

808,413:

___

___

TRY these:

a. Write in standard notation:

 a. Three billion, four hundred fifty–two thousand, eight hundred fifty

 ____

 b. Eighty–six thousand four hundred twelve

 ____

b. Write in expanded notation:

 a. 34, 076

 ____

 ____

 b. 570, 003

 ____

 ____

Ordering of Whole Numbers

Graph the following numbers: 1, 3, 5 and 7, then compare them using the inequality signs, < (less than) and > (greater than).

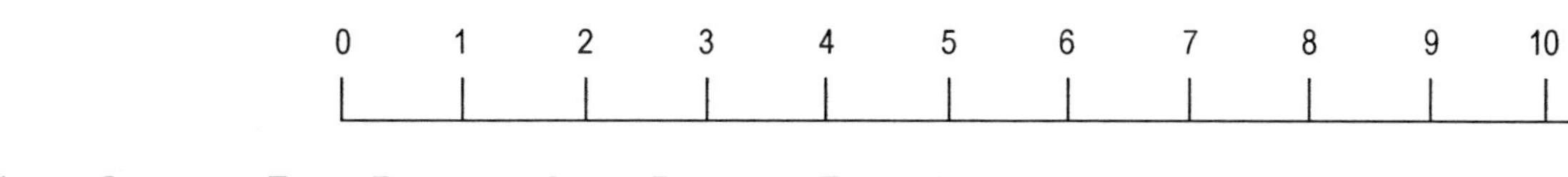

1____3 7____5 3____5 7____1

Graphing an Inequality on a Number line

When graphing inequalities on a number line, you must first identify is if it is an open or closed statement.

An **open statement** is represented by an open circle at the specific number in the statement because the problem has a < or > represented. You can associate the phrase "less than or greater than" with the problem.

A **closed statement** is represented by a solid circle at the specific number in the statement because the problem has a ≤ or ≥ represented. You can associate the phrase "less than or equal to" and "greater than or equal to" with the problem.

EXAMPLES:

Graph numbers less than 3

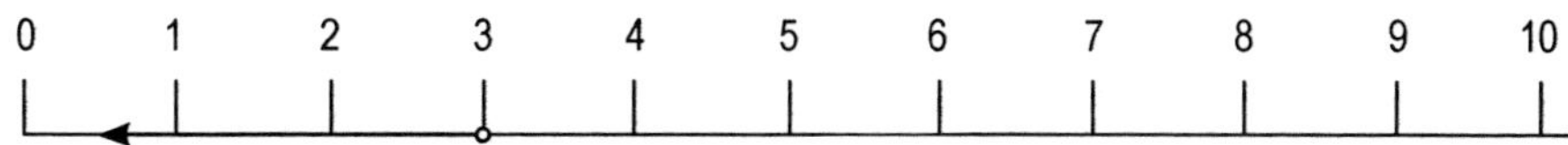

Graph numbers greater than or equal to 2

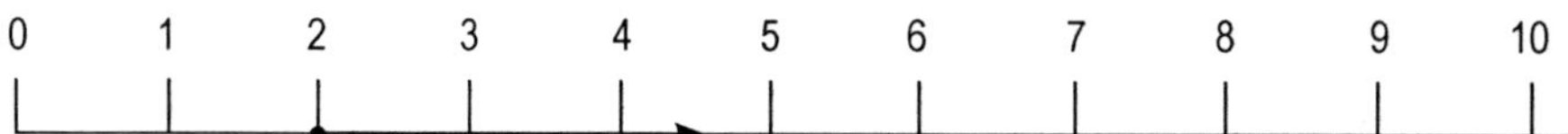

PRACTICE:

Graph numbers greater than 1

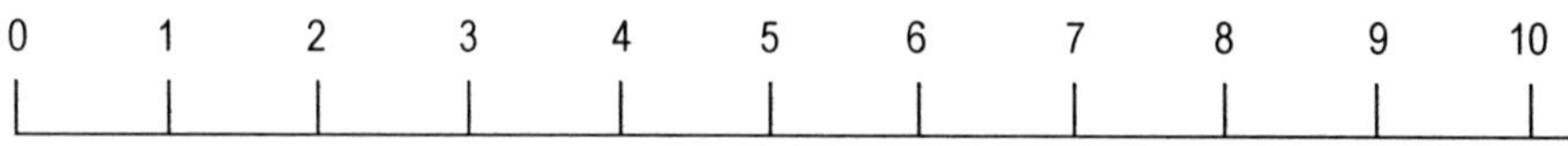

Graph numbers less than or equal to 4

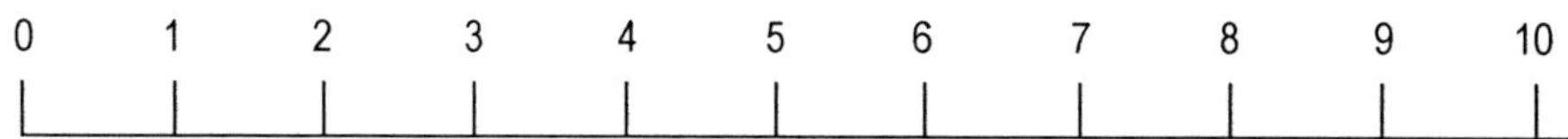

Rounding Whole Numbers

If the **test digit** ≥5, you will round up by increasing the rounding digit by 1

If the **test digit** < 5, you will round down by leaving the rounding digit unchanged.

First you look at the tens place; a suggestion is to underline the place you want and then circle the digit after it. If the circled digit is 5 or greater, the underlined digit will increase by a value of one and the circled digit becomes zero. If the circled digit is less than 5, then the underlined digit stays as is and replace the circled digit with a zero.

Ex: Round 123 to the nearest ten.

1<u>2</u>③ since the 3 is less than 5 it becomes 120

Ex: Round 1256 to the nearest hundreds.

1 <u>2</u> ⑤ 6 since it is a 5 or greater it becomes 1300

EXAMPLE:

1. Round each number to the nearest ten:

   a. 45,743: __________

   b. 4,277: __________

   C. 304: __________

2. Round each number to the nearest hundred:

   a. 378,233: __________

   b. 6,789: __________

3. Round to the nearest thousand:

   a. 32,600: __________

   b. 73,008: __________

4. Round to the nearest ten thousand:

   126,424: ____________

   15,985: ______________

Constructing and Interpreting Graphs

1. First collect your data in a chart format like the one provided.

2. Label your graph/charts vertical and horizontal axis with what each represents and title your graph/chart with what it is representing.

Name	Age

Age	Frequency
11–20	
21–30	
31–40	
41–50	
51–60	
61–70	

EXTRA PRACTICE:

1. Graph 0, 2, 4, 6, and 8.

0 1 2 3 4 5 6 7 8 9 10

2. Graph whole numbers less than 5.

0 1 2 3 4 5 6 7 8 9 10

3. Graph whole numbers greater than or equal to 3

0 1 2 3 4 5 6 7 8 9 10

Place an > or < symbol to make a true statement.

4. 309____300

5. 3____67

6. 2,502______2,502

Write each number in expanded notation and then in word form.

7. 508: ____________________

____

8. 73,009: ____________________

____

9. 3,609: ____________________

____

Write each number in standard notation.

10. 2 thousands + 7 hundreds + 3 tens + 6 ones:____________________

11. Twenty–seven thousand five hundred ninety–eight: ____________________

12. Eighty–six thousand four hundred twelve: ____________________

Round 79,593 to the nearest:

13. Ten: ______________________________

14. Thousand: ______________________________

15. Hundred: ______________________________

Round $419, 161 to the nearest:

16. $10: ______________________________

17. $100: ______________________________

18. $10,000: ______________________________

Applications:

The speed of light in a vacuum is 299,792,458 meters per second. Round to the nearest hundred thousand meters per second. ______________________________

HOMEWORK ASSIGNMENT: 1.1

Name: ______________________________

Write the following in expanded notation:

1. 3,960: ______________________________

2. 73, 009: ______________________________

____

Write the following in standard notation:

3. 4 hundreds + 2 tens + 5 ones:

____

4. 2 thousands + 7 hundreds + 3 tens + 6 ones:

____

5. Three thousand seven hundred thirty–six:

____

6. Seven million, four hundred fifty–two thousand, eight hundred sixty–seven:

____

7. Eighty–five thousand seven hundred sixteen:

____

Round 81, 253 to the nearest…

8. Ten: ______________________________

9. Hundred: ______________________________

10. Thousand: ______________________________

11. Ten–thousand: ______________________________

Compare using < or > symbol.

12. 51________ 65

13. 312________ 311

14. 47________ 53

15. 99________ 98

1.2 ADDING AND SUBTRACTING WHOLE NUMBERS

Objectives:

- Add and subtract whole numbers.
- Multiply and Divide whole numbers.
- Find the perimeter and area of a polygon.

Properties of Addition and Multiplication

Commutative Property: if a and b represent numbers, then

$a + b = b + a$

$a \times b = b \times a$

EXAMPLES:

a. $5 + 3 = 3 + 5$ b. $5 \times 3 = 3 \times 5$

PRACTICE:

a. $2 + 7 =$ b. $4 + 9 =$

c. $2 \times 5 =$ d. $3 \times 7 =$

Associative property: if a and b represent numbers, then

$(a + b) + c = a + (b + c)$

$(a \times b) \times c = a \times (b \times c)$

EXAMPLE:

a. $(3 + 5) + 6 = 3 + (5 + 6)$ b. $(3 \times 5) \times 6 = 3 \times (5 \times 6)$

PRACTICE:

a. $(2 + 4) + 5 =$ b. $(4 + 7) + 9 =$

c. $(2 \times 3) \times 4 =$ d. $(4 \times 5) \times 7 =$

Identity Property

If 0 is added to any number, the sum will be the same number.

$a + 0 = a$

EXAMPLE:

$4 + 0 =$

If 1 is multiplied by any number, the product will be the same number.

$a \times 1 = a$

EXAMPLE:

$4 \times 1 =$

Adding Whole Numbers

Terms: the individual parts, such as: 3, 5, and 10

Sum: the combination of the individual terms

EXAMPLES:

a. $\begin{array}{r} 75 \\ +16 \\ \hline \end{array}$

b. $\begin{array}{r} 1{,}729 \\ +873 \\ \hline \end{array}$

c. $\begin{array}{r} 1{,}385 \\ 246 \\ 139 \\ +10 \\ \hline \end{array}$

Subtracting Whole Numbers

Not commutative or associative like addition, when grouped differently the result is always different.

EXAMPLES:

a. $\begin{array}{r} 43 \\ -15 \\ \hline \end{array}$

b. $\begin{array}{r} 7{,}941 \\ -738 \\ \hline \end{array}$

c. $\begin{array}{r} 1{,}202 \\ -346 \\ \hline \end{array}$

FACT Families: Addition and Subtraction

A **fact family** is a group of four statements each of which represent the same meaning. All four statements have the same meanings. Fact families contain 3 numbers and 2 types of operations. There are 2 types of fact families: addition and subtraction fact families; multiplication and division fact families.

FOR EXAMPLE:

3, 4, *and* 7

$3 + 4 = 7$

$4 + 3 = 7$

$7 - 3 = 4$

$7 - 4 = 3$

There are several ways to represent fact families and they are in the proceeding templates. The easiest way to use them is with a clear overhead sheet and dry erase markers so that they can be re-used.

How to Use the Templates

A fact family consists of three numbers. Just like any family the members, or numbers, are related and there is always at least four math facts to be made with them. For example, look at these members of a fact family: 3, 7, and 10.

The Fact Family Relationships

In the family 3, 7, and 10 they are related because you can add two of the numbers to get the last number.

- $3 + 7 = 10$

You can also switch the first two numbers, using the commutative property of addition, and still get the same answer.

- $7 + 3 = 10$

Fact Family Cousins

If addition is the direct relationship between these family members, then subtraction is the family cousin through the inverse property. Subtraction is the opposite of addition, but it's still related. The problems still only use the three members of the family.

- $10 - 7 = 3$
- $10 - 3 = 7$

Keeping Track of All Family Members

Once you know the relationships of the fact family members, it's easy to see who is missing at a quick glance. Solving addition and subtraction problems is then much easier and starts to become automatic. Take for example, this problem:

- $7 + \_\_\_\_ = 10$

You should quickly be able to recognize 3 as the missing family member.

Dominoes Fact Family Template

______ + ______ = ______

______ + ______ = ______

______ − ______ = ______

______ − ______ = ______

Triangles Fact Family Template

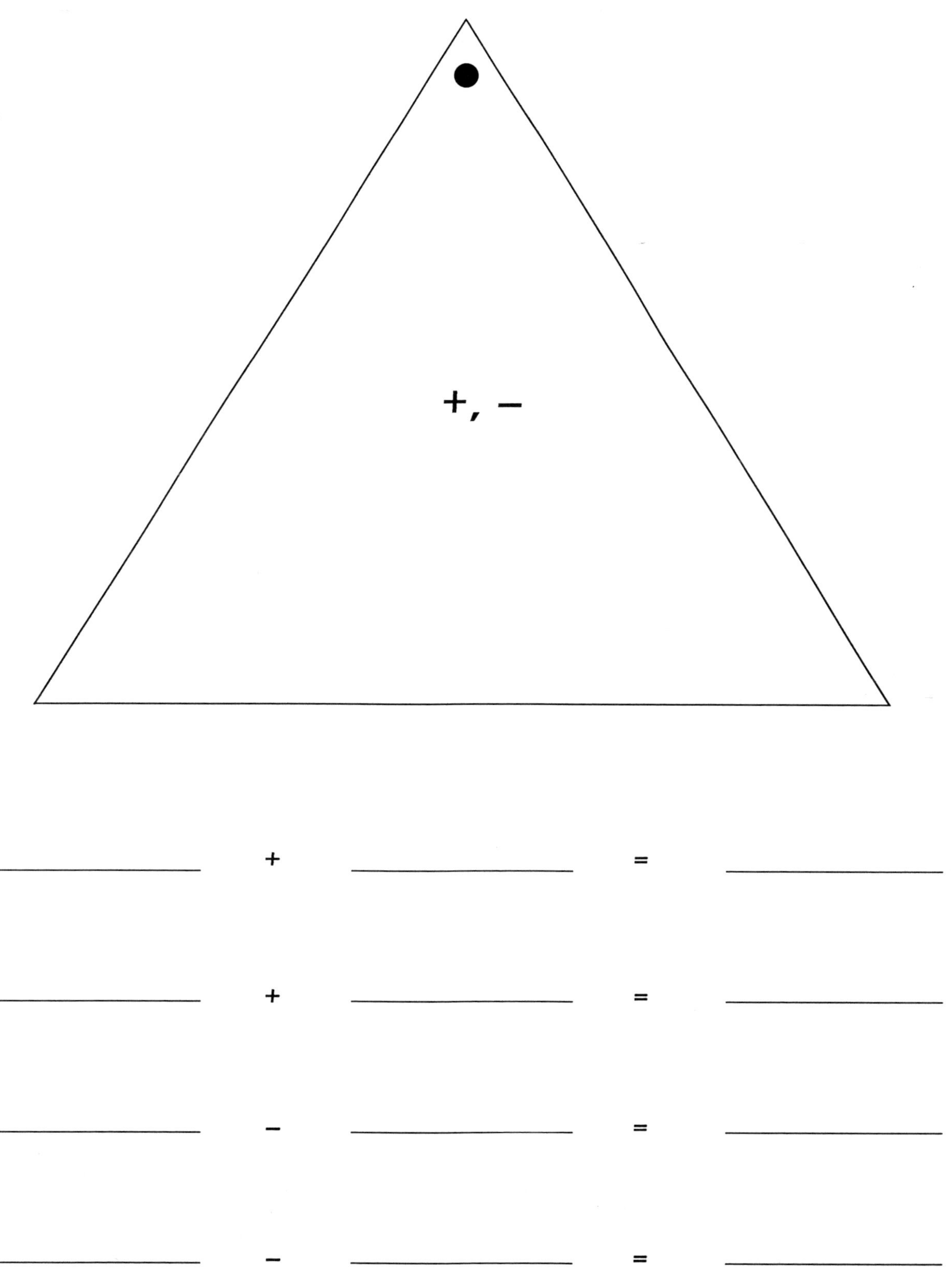

House Fact Family Template

__________ + __________ = __________

__________ + __________ = __________

__________ – __________ = __________

__________ – __________ = __________

PRACTICE:

1.

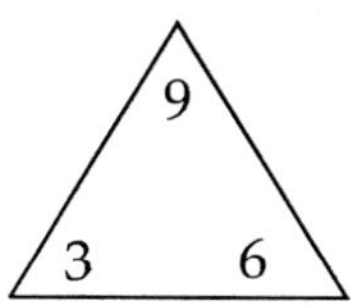

□ + □ = □

□ + □ = □

□ − □ = □

□ − □ = □

2.

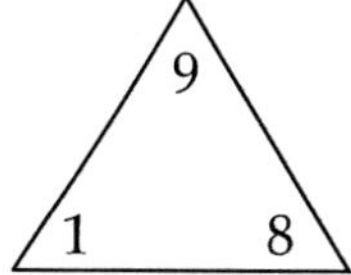

□ + □ = □

□ + □ = □

□ − □ = □

□ − □ = □

3.

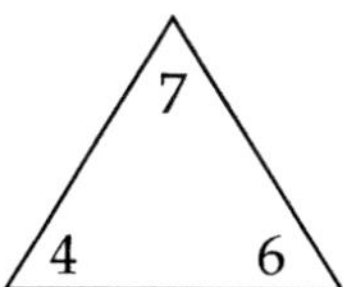

□ + □ = □

□ + □ = □

□ − □ = □

□ − □ = □

4.

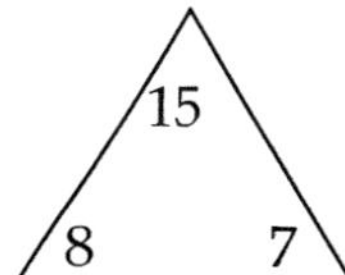

□ + □ = □

□ + □ = □

□ − □ = □

□ − □ = □

5.

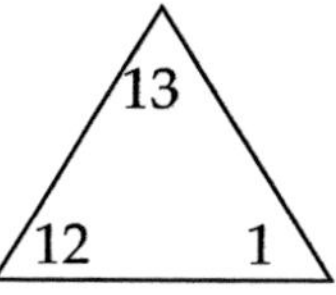

□ + □ = □

□ + □ = □

□ − □ = □

□ − □ = □

6.

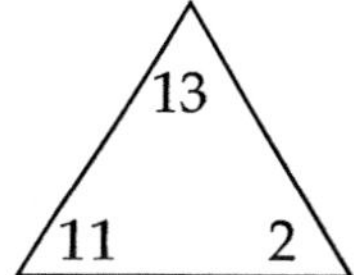

□ + □ = □

□ + □ = □

□ − □ = □

□ − □ = □

1.3 MULTIPLYING AND DIVIDING WHOLE NUMBERS

Multiplying Whole Numbers

Product: the result of multiplying two **factors**, the numbers being multiplied.

Symbols: x, ·, ( )

EXAMPLES:

a. $\begin{array}{r} 56 \\ \underline{\times 9} \end{array}$

b. $\begin{array}{r} 19 \\ \underline{\times 14} \end{array}$

c. $\begin{array}{r} 165 \\ \underline{\times 3} \end{array}$

Dividing Whole Numbers

Dividend: the number that is being divided

Divisor: the number that is being used to divide

Quotient: the result (answer)

Remainder: division with 0, division with 1, same number dividing.

Symbols: ÷, –, $\overline{)\ }$, /

Division with 0

1. If a represents any nonzero number, $\frac{0}{a} = 0$.
2. If a represents any nonzero number, $\frac{a}{0}$ is undefined.
3. $\frac{0}{0}$ is undetermined.

Division Properties

If a represents any number, (provided that a)

EXAMPLES:

a. 832 ÷ 23

b. 1,990 ÷ 165

c. $39\overline{)7995}$

d. $57\overline{)1795}$

FACT Families: Multiplication and Division

The multiplication table is a great tool for exploring number patterns and relationships. Multiplication tables are also useful for showing division concepts and fact families.

Look at the multiplication table below.

The numbers along the left side and top are factors. The numbers inside are products.

factors ↓

factors →

X	0	1	2	3	4	5	6	7	8	9	10
0	0	0	0	0	0	0	0	0	0	0	0
1	0	1	2	3	4	5	6	7	8	9	10
2	0	2	4	6	8	10	12	14	16	18	20
3	0	3	6	9	12	15	18	21	24	27	30
4	0	4	8	12	16	20	24	28	32	36	40
5	0	5	10	15	20	25	30	35	40	45	50
6	0	6	12	18	24	30	36	42	48	54	60
7	0	7	14	21	28	35	42	49	56	63	70
8	0	8	16	24	32	40	48	56	64	72	80
9	0	9	18	27	36	45	54	63	72	81	90
10	0	10	20	30	40	50	60	70	80	90	100

A **fact family** is a group of four statements each of which represent the same meaning. All four statements have the same meanings. Fact families contain 3 numbers and 2 types of operations. There are 2 types of fact families: addition and subtraction fact families; multiplication and division fact families.

FOR EXAMPLE:

3, 5 and 15

$3 \cdot 5 = 15$

$5 \cdot 3 = 15$

$15 \div 3 = 15$

$15 \div 5 = 3$

Perfect Squares	Square roots
$1^2 = 1$	$\sqrt{1} = 1$
$2^2 = 4$	$\sqrt{4} = \sqrt{2*2} = 2$
$3^2 = 9$	$\sqrt{9} = \sqrt{3*3} = 3$
$4^2 = 16$	$\sqrt{16} = \sqrt{4*4} = 4$
$5^2 = 25$	$\sqrt{25} = \sqrt{5*5} = 5$
$6^2 = 36$	$\sqrt{36} = \sqrt{6*6} = 6$
$7^2 = 49$	$\sqrt{49} = \sqrt{7*7} = 7$
$8^2 = 64$	$\sqrt{64} = \sqrt{8*8} = 8$
$9^2 = 81$	$\sqrt{81} = \sqrt{9*9} = 9$
$10^2 = 100$	$\sqrt{100} = \sqrt{10*10} = 10$
$11^2 = 121$	$\sqrt{121} = \sqrt{11*11} = 11$
$12^2 = 144$	$\sqrt{144} = \sqrt{12*12} = 12$
$13^2 = 169$	$\sqrt{169} = \sqrt{13*13} = 13$

How to Use the Templates

A fact family consists of three numbers. Just like any family the members, or numbers, are related and there is always at least four math facts to be made with them. For example, look at these members of a fact family: 3, 5, and 15.

The Fact Family Relationships

In the family 3, 5, and 15 they are related because you can multiply two of the numbers to get the last number.

- $3 \cdot 5 = 15$

You can also switch the first two numbers, using the commutative property of multiplication, and still get the same answer.

- $5 \cdot 3 = 15$

Fact Family Cousins

Ifmultipllication is the direct relationship between these family members, thendivision is the family cousin through the inverse property. Division is the opposite of multiplication, but it's still related. The problems still only use the three members of the family.

- $\frac{15}{3} = 5$
- $\frac{15}{3} = 3$

Keeping Track of All Family Members

Once you know the relationships of the fact family members, it's easy to see who is missing at a quick glance. Solving multiplication and division problems is then much easier and starts to become automatic. Take for example, this problem:

$3 \cdot ? = 15$ $\qquad$ $\frac{15}{?} = 3$

You should quickly be able to recognize 5 as the missing family member.

Triangle Fact Family Template

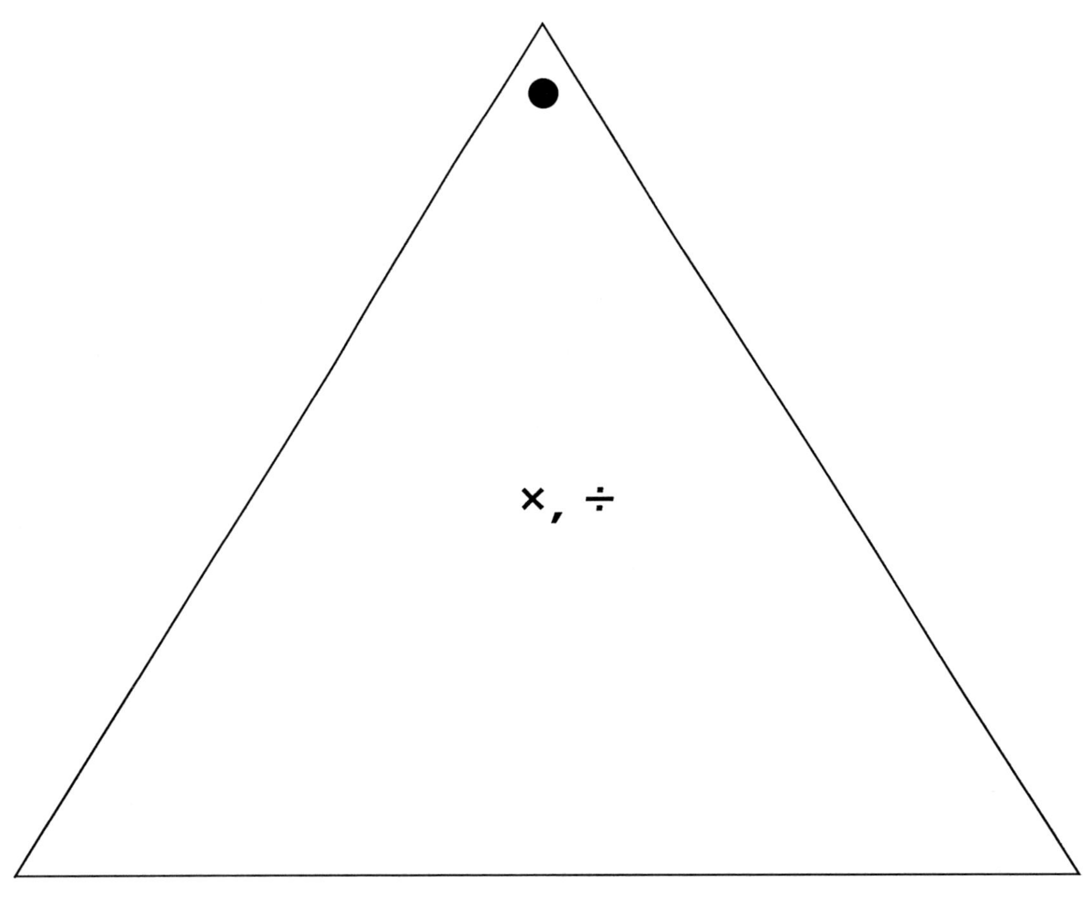

__________ × __________ = __________

__________ × __________ = __________

__________ ÷ __________ = __________

__________ ÷ __________ = __________

PRACTICE:

1.

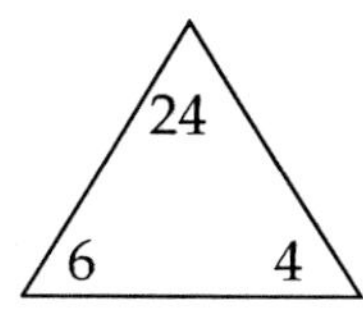

☐ × ☐ = ☐
☐ × ☐ = ☐
☐ ÷ ☐ = ☐
☐ ÷ ☐ = ☐

2.

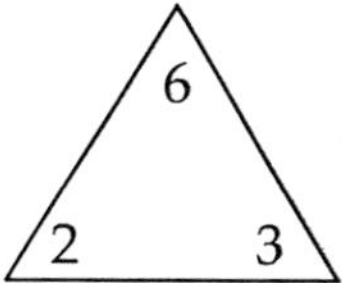

☐ × ☐ = ☐
☐ × ☐ = ☐
☐ ÷ ☐ = ☐
☐ ÷ ☐ = ☐

3.

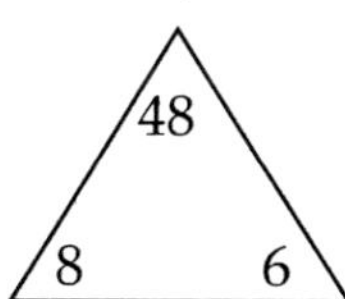

☐ × ☐ = ☐
☐ × ☐ = ☐
☐ ÷ ☐ = ☐
☐ ÷ ☐ = ☐

4.

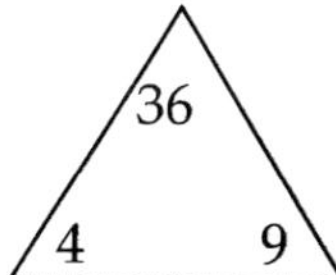

☐ × ☐ = ☐
☐ × ☐ = ☐
☐ ÷ ☐ = ☐
☐ ÷ ☐ = ☐

5.

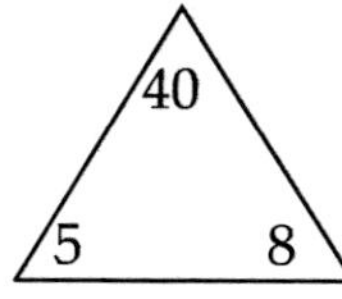

☐ × ☐ = ☐
☐ × ☐ = ☐
☐ ÷ ☐ = ☐
☐ ÷ ☐ = ☐

6.

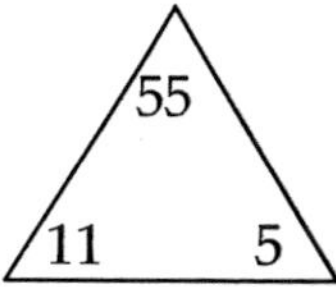

☐ × ☐ = ☐
☐ × ☐ = ☐
☐ ÷ ☐ = ☐
☐ ÷ ☐ = ☐

Perimeter and Area of a rectangle and a square

Rectangle: 4–sided figure that has 4 right angles and 2 pairs of equal sides.

Square: 4–sided figure that has 4 right angles and 4 equal sides.

Length: the longest side of a rectangle.

Width: the shortest side of a rectangle.

Dimensions: the length and width of a rectangle together; ie.length by width.

Perimeter: distance around a figure with three or more sides; i.e. Triangle, square, and rectangle to name a few.

Formulas:

Perimeter of a rectangle

P= length + length + width + width → $P = 2l + 2w$

Area of a rectangle

A= length × width → $A = lw$

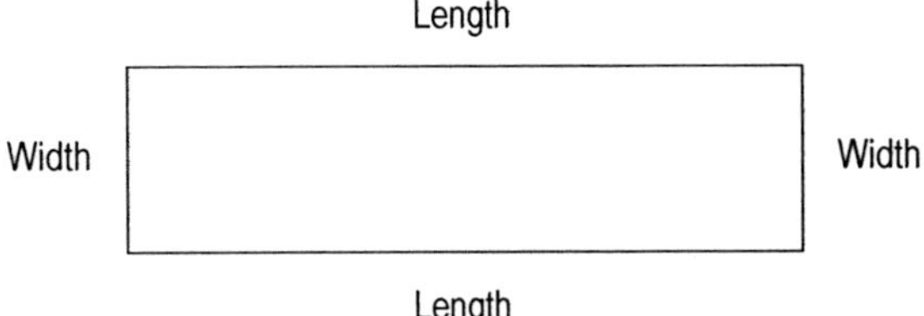

EXAMPLE:

Find the perimeter and area of a dollar bill with dimensions 65mm and 156mm.(Hint: Draw a picture and label the sides.)

Perimeter of a square

P = side + side + side + side → $P = 4s$

Area of a square

A= side × side → $A = s^2$

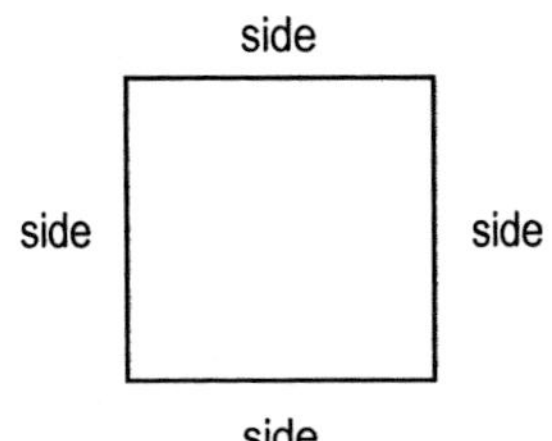

EXAMPLE:

Find the area and perimeter of a square plate with a length of 6in.

(Hint: Draw a picture and label the sides.)

Distance Formula

This formula can be rearranged in three ways to find a particular piece.

If you are looking for distance you would use the following:

Distance = (rate)(time) → $D = rt$

If looking for rate you would use the following:

Rate = Distance/time → $r = \frac{D}{t}$

If looking for time you would use the following:

Time = Distance/rate → $t = \frac{D}{r}$

PRACTICE:

1. Find the distance traveled if a vehicle is moving at a rate of 45mph and was on the road for 3 hours.

2. What is the rate of a plane that had traveled a distance of 2000 kilometers in a time frame of 4 hours?

3. How much time did it take a train to travel 500 miles at a rate of 50 mph?

EXTRA PRACTICE:

Perform each addition:

1. (95 + 16) + 39 = ____________________

2. 3,567 + 8,778 = ____________________

3. 3,156 + 1,578 + 578 = ____________________

Find the perimeter of the following:

4.

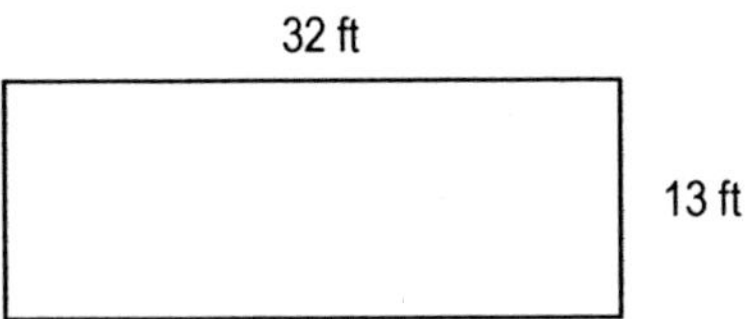

5.

5 yd

5 yd

5 yd

Perform each subtraction:

6. 633 – (598 – 30) = ____________________

7. 498 – 17 – 162 = ____________________

8. 35, 201 – 23, 399 = ____________________

Perform the following:

9. $43 - 12 + 9 =$ ____________________

10. $600 + 99 - 54 =$ ____________________

11. $59 - 16 + 2 =$ ____________________

Applications:

12. For a 17–mile trip, Wanda paid the taxi driver $23. If $5 was a tip, how much was the fare?

13. Gold melts at about 1,947° F. The melting point of silver is 183° F lower. What is the melting point of silver?

14. A student wants to make the 2,221–mile trip from Detroit to Seattle in three days. If she drives 751 miles on the first day and 875 miles on the second day, how far must she travel on the third day?

Perform each multiplication:

15. $9 \cdot (4 \cdot 5) =$ ____________________

16. $73 \times 59 =$ ____________________

17. $123{,}112 \times 46 =$ ____________________

Find the area of the following:

18.

50 m

2 m

19.

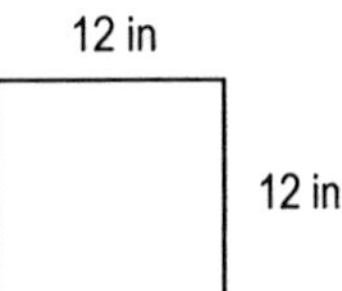

Perform each division:

20. $132 \div 11 =$ ________________________

21. $13\overline{)949} =$ ________________________

22. $83\overline{)3{,}280} =$ ________________________

Applications:

23. A cook worked 12 hours at $11 per hour. How much did she earn?

24. Mia owns an apartment building with 18 units. Each unit generates a monthly income of $450. Find her total monthly income. Find her total yearly income.

25. A cereal maker advertises "Two cups of raisins in every box." What is the number of cups of raisins in a case of 36 boxes of cereal?

26. A first–grade class received 73 half–pint cartons of milk to distribute evenly among 23 students. How many cartons were left over?

27. A touring rock group travels in a bus that has a range of 700 miles on one tank (140 gallons) of gasoline. How far can the bus travel on 1 gallon of gas?

28. How many feet more than 2 miles is 11,000 feet? (Hint: 5,280 feet = 1 mile)

29. Which has the greater area: a rectangular room that is 14ft by 17ft or a square room that is 16 ft on each side? Which has the greater perimeter?

30. A rectangular garden is 27ft long and 19ft wide. A path in the garden uses 125ft^2 of space. How many square feet are left for planting?

HOMEWORK ASSIGNMENT: 1.2 -1.3

Name: ______________________________

Add:

1. (95 + 16) + 39

2. (4,321 + 213) + 5,234

3. 1,246 + (5,799 + 6,889) + 29

Subtract:

4. 162 – 17 – 9

5. 7,346 – 3, 298

6. Subtract 270 from 434

Find the perimeter of the following:

7.

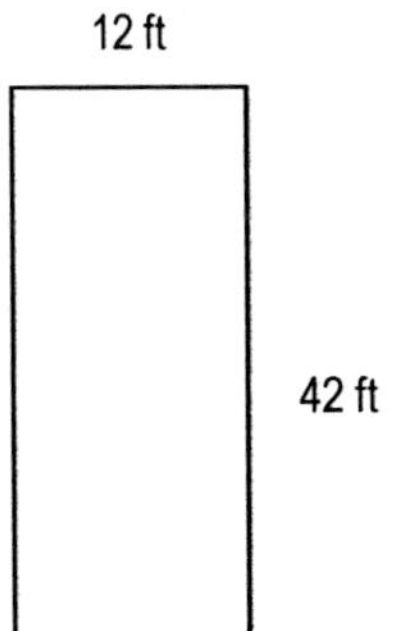

8.

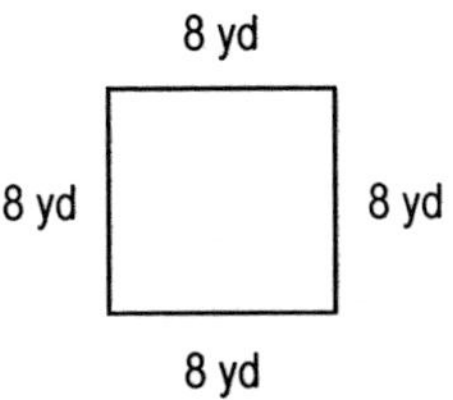

9. For a 23–mile trip from Orlando International Airport to her apartment, Felisha paid the taxi driver $29. If $5 was a tip, how much was the fare?

10. Astronaut Walter Schirra's first space flight orbited the Earth 6 times and lasted 9 hours. His second flight orbited the Earth 16 times and lasted 26 hours. How long was Schirra in space?

11. A savings account contained $523. After a deposit of $54 and a withdrawal of $239, how much is still in the account?

12. A group of college students want to make the 4,405–mile trip from Miami to Los Angeles in 5 days. If they drive 1,112 miles the 1st day, 523 miles on the 2nd day, 634 miles on the 3rd day and 1123 miles on the 4th day, how far must they travel on the 5th day?

Multiply:

13. $(3 \cdot 5) \cdot 12$

14. $73 \cdot 59$

15. $232 \cdot 53$

16. $123{,}112 \cdot 43$

Divide:

17. $132 \div 11$

18. $33\overline{)1{,}353}$

19. $42\overline{)1{,}273}$

20. $99\overline{)9{,}876}$

Find the area of the following:

21.

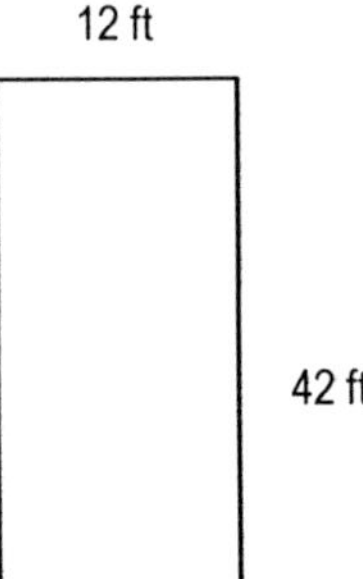

22.

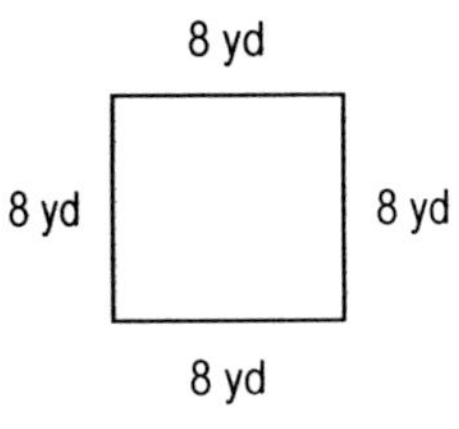

23. A hostess at I.H.O.P. worked for 25 hours at $12.50 per hour. How much did she earn?

24. Michael owns an Orlando apartment building with 22 units. Each unit generates a monthly income of $650. Find his total monthly income.

25. A first–grade class in Kissimmee received 75 half–pint cartons of milk to distribute evenly among 23 students. How many cartons were left over?

26. A touring rock group in Florida travels in a bus that has a range of 650 miles on one tank (130 gallons) of gasoline. How far can the bus travel on 1 gallon of gas? We are finding the miles per gallon.

1.4 PRIME FACTORS AND EXPONENTS

- Breakdown whole numbers to its prime number.
- Write numbers in exponential form.

Even and Odd Whole Numbers

Even: divisible by 2

Odd: not divisible by 2

Prime Numbers

It is a whole number, greater than 1, that has only 1 and itself as a factor; 2, 3, 5, 7, 11, 13, 17, 19… to name a few.

Composite Numbers

There are whole numbers, greater than 1, that are not prime; 4, 6, 8, 9, 10, 12, 14, 15, 16,… to name a few.

Factoring a Whole Number

Factor: numbers that are multiplied together to get a final product.

Factoring: show a number as a product of other whole numbers.

EXAMPLE:

Find the factors of:

20 23 72

100 81 75

Prime Factorization:

write a whole number as the product of only prime numbers.

EXAMPLE:

Find the prime factorization of:

72	210	150
$8 \cdot 9$		
$2 \cdot 4 \cdot 3 \cdot 3$		
$2 \cdot 2 \cdot 2 \cdot 3 \cdot 3$	$2^2 \cdot 3^2$	
400	126	243

Exponents:

Exponent: small superscript number indicating repeated multiplication of the same number.

Base: represents a factor of a larger number.

EXAMPLE:

Write as a power:

a. $2 \times 2 \times 2 \times 2 \times 2 \times 2$

b. $7 \times 7 \times 7 \times 7$

EXAMPLE:

Find each power:

a. 5^2

b. 2^7

c. 8^3

Your turn:

The prime factorization of a number is $2^3 \times 3^2 \times 5$. What is the number?

The prime factorization of a number is. What is the number?

EXTRA PRACTICE:

1. Write 30 as the product of three factors.

2. Find the factors of 24.

3. Find the factors of 100.

Prime factorization of a whole number is given. Find the number.

4. $2 \times 3 \times 3 \times 5$

5. $3^3 \times 2$

6. $11^2 \times 5$

7. Find the factors of 30 and 165. What prime factors do they have in common?

Write the following in notation form:

8. $2^2 \cdot 5^3 \cdot 11^2$

9. $3^3 \cdot 5^2 \cdot 7$

Write the number in prime–factored form.

10. 220

11. 243

Find the end product.

12. $3^2(2^3)$

13. $23^2 \cdot 13^2$

Applications:

14. After one hour, a cell has divided to form another cell. In another hour, these two cells have divided so that four cells exist. In another hour, these four cells divide so that eight exist.

    a. How many cells exist at the end of the fourth hour?

    b. The number of cells that exist after each division can be found using an exponential expression. What is the base?

    c. Find the number of cells after the seventh hour.

15. When a university band lines up in eight rows of 15 musicians, there are five musicians left over. How many band members are there?

HOMEWORK ASSIGNMENT: 1.4

Name: ______________________________

Write in prime–factored form:

1. 100

2. 441

3. 126

4. 162

5. 128

6. 300

7. 136

8. Find the factors of 30 and 242. What prime factors do they have in common?

Find the end product.

9. $3^2 \cdot 4^3 \cdot 5^2$

10. After one hour, a cell has divided to form another cell. In another hour, these two cells have divided so that four cells exist. In another hour, these four cells divide so that eight exist.

 a. How many cells exist at the end of the sixth hour?

 b. The number of cells that exist after each division can be found using an exponential expression. What is the base?

1.5 ORDER OF OPERATIONS

G (grouping: parentheses ( ), brackets [ ], absolute value bars | |, braces { })

E (exponents)

MD (multiply, divide) → whichever comes first left to right
→

AS (add, subtract) → whichever comes first left to right
→

$2 + 3 \cdot 4 \div 2 + 7$

Since there are no grouping symbols given, you must:

1. Multiply 3 and 4
2. Then divide by 2
3. Then add from the left to the right

$2 + 3 \cdot 4 \div 2 + 7$

$$2 + 12 \div 2 + 7$$
$$2 + 6 + 7$$
$$15$$

EXAMPLES:

1. $4 \times 3^3 - 6$
2. $10 - 2 \cdot 3 + 24$
3. $36 \div 9 + 4(2)(3)$
4. $(1 + 3)^4$
5. $1 + 3^4$
6. $50 - 4(12 - 5 \cdot 2)$

7. $16 + 2[14 - 3(5 - 2)]$

8. $140 - 7[4 + 3(6 - 2)]$ 9)3(14) – 6

9. $\dfrac{3(14) - 6}{2(3^2)}$

Average or $\dfrac{M + E + A + N}{4}$

EXAMPLES:

1. Syracuse University won the 2003 NCAA men's basketball championship. Find their average margin of victory in their six tournament games, which they won by 11, 12, 1, 16, 11, and 3 points.

2. The table shows the sandwiches Subway advertises as its low–fat menu. What is the average (mean) number of calories for the group of sandwiches?

6–inch Sub	Calories	Fat (g)
Veggie Delite	237	3
Turkey Breast	289	4
Turkey Breast & Ham	295	5
Ham	302	5
Roast Beef	303	5
Subway Club	312	5
Roasted Chicken Breast	348	6

EXTRA PRACTICE:

1. $4 \cdot 2 + 2 \cdot 4$

2. $3 \cdot 2 \cdot 3^4 \cdot 5$

3. $5 \cdot 10^3 + 2 \cdot 10^2 + 3 \cdot 10 + 9$

4. $7 + \left(5^3 - \dfrac{200}{2}\right)$

5. $4[50 - (3^3 - 5^2)]$

6. $\dfrac{(4^3 - 2) + 7}{5(2 + 4) - 7}$

7. $\dfrac{25 - (2 \cdot 3 - 1)}{2 \cdot 9 - 8}$

Applications:

8. At the supermarket, Carlos has 2 cases of soda, 4 bags of potato chips, and 2 cans of dip in his cart. Each case of soda costs $6, each bag of chips cost $2,and each can of dip costs $1. Find the total cost of the groceries.

9. In a psychology class, a student had test scores of 94, 85, 81, 77, and 89. He also overslept, missed the final exam, and received a 0 on it. What was his test average in the class?

HOMEWORK ASSIGNMENT: 1.5

Name: ____________________

Evaluate:

1. $6 + 2\,(5 + 4)$

2. $(9 - 5)^3 + 8$

3. $7 + \left(5^3 - \frac{200}{2}\right)$

4. $15 - \frac{24}{6} + 8 \cdot 2$

5. $4[50 - (3^3 - 5^2)]$

6. $15 + 5[12 - (2^2 + 4)]$

7. $8[6(6) - 6^2] + 4(5)$

8. $\frac{5^2 + 17}{6 - 2^2}$

9. $\frac{25 - (2 \cdot 3 - 1)}{2 \cdot 9 - 8}$

10. $\frac{(5-3)^2 + 2}{4^2 - (8+2)}$

11. One week in December, the temperatures in Orlando, Florida, were 75°, 80°, 63°, 77°, 69°, 53°, and 72°. Find the week's average (mean) temperature.

12. In a psychology class, a student had test scores of 94, 85, 80, and 79. He also overslept, missed the final exam, and received a zero on it. What was his test average in the class?

CHAPTER 2

INTEGERS

2.1 AN INTRODUCTION TO THE INTEGERS

Objectives:

- Use integers to represent real life situations.
- Graph on a number line.
- Compare numbers.
- Find the opposite of a given integer.

The set of natural numbers (does not contain the value of zero)

The set of whole numbers (contains the value of zero)

The set of integers (contains both positive and negative numbers)

The set of rational numbers includes the naturals, whole, and integer sets.

The set of real numbers contains the naturals, whole, integers, rational, and irrational numbers.

Classify the following numbers:

a. –3

b. 0

c. 4

New Number Line

Graph and compare the following numbers: 3, 1.5, –2, –1.5, and 0

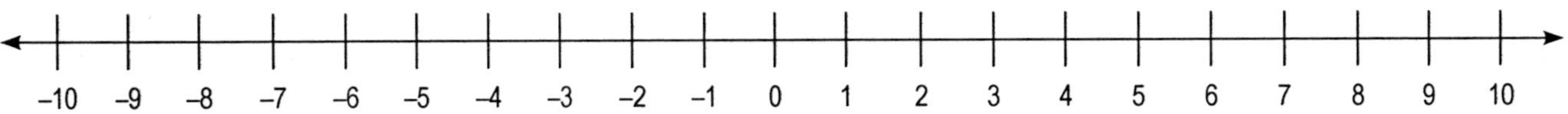

Absolute value: the distance on a number line between the number and zero; always a positive value, never negative.

The opposite of a number: two numbers that are the same distance from 0 on the number line, but are located on opposite sides of 0.

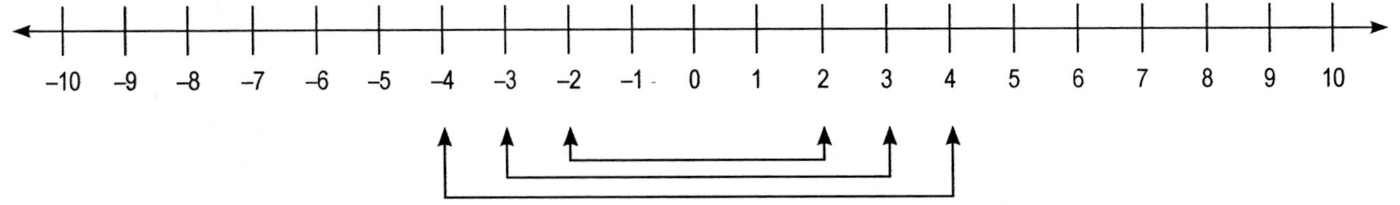

Double Negative Rule

If is any number, then

EXAMPLE:

Simplify each expression:

EXTRA PRACTICE:

1. On a number line, what number is 3 units to the right of –7?

2. Name two numbers on a number line that are a distance of 5 away form –3.

3. Translate each phrase to mathematical symbols.

   a. The opposite of negative 8

   b. Absolute value of negative eight

   c. Eight minus eight

   d. Opposite of the absolute value of negative eight

Simplify

4. $|-8|$

5. $-|-6|$

6. $-(-4)$

Graph

7. {–3,0,3,–2}

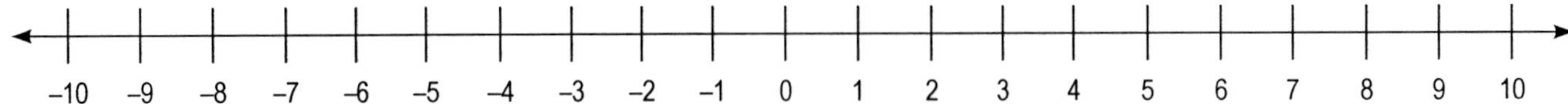

8. {–4,–2,1,3}

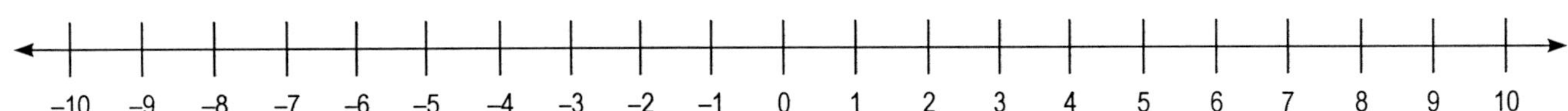

9. Graph the opposite of –3,the opposite of 4,and the absolute value of –1.

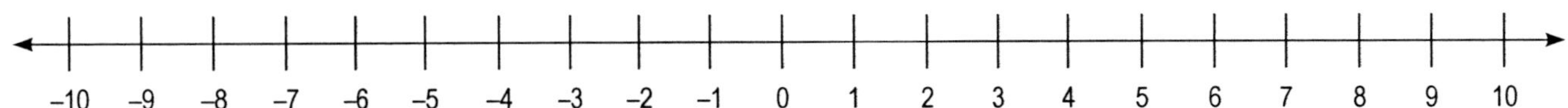

10. A week of daily reports listing the height of a river in comparison to flood stage is given by the following table. Create a bar graph using the given information.

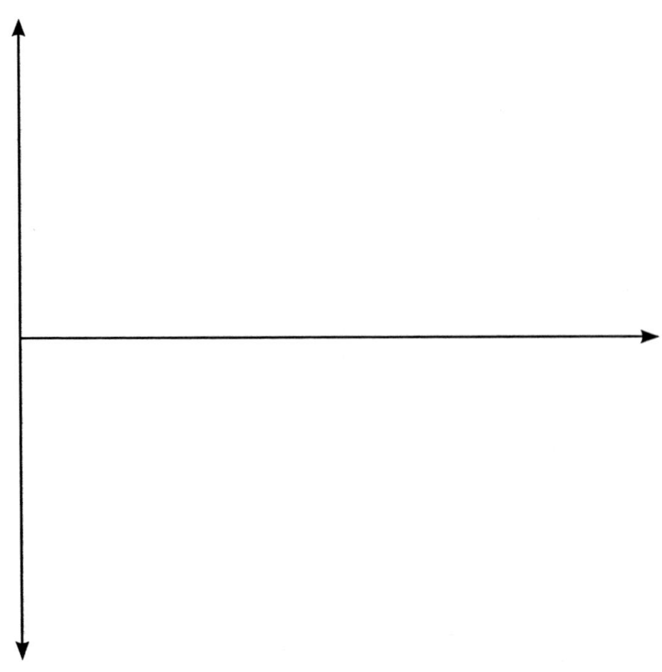

Flood stage report	
Sun.	2ft over
Mon.	3ft below
Tues.	4ft below
Wed.	2ft below
Thurs.	1ft over
Fri.	3ft over
Sat.	4ft over

HOMEWORK ASSIGNMENT: 2.1

Name: ______________________________

1. Name two numbers that are a distance of 5 away from –3.

2. Name two numbers that are a distance of 4 away from 2.

Graph each set of numbers.

3. {–3.5, –.5, 1.5, 3, –2.5}

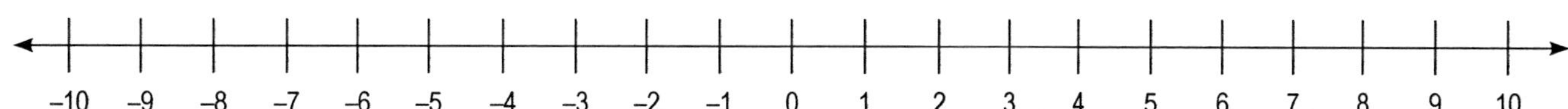

4. The absolute value of 3, the opposite of 3, and the number that is 1 less than –3.

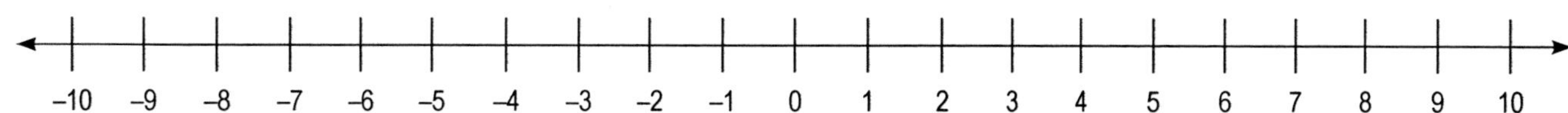

5. Graph numbers less than –2. Remember properties of inequalities.

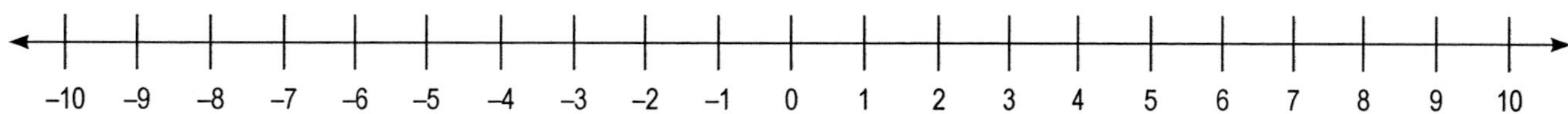

6. Graph numbers greater than or equal to –9. Remember properties of inequalities.

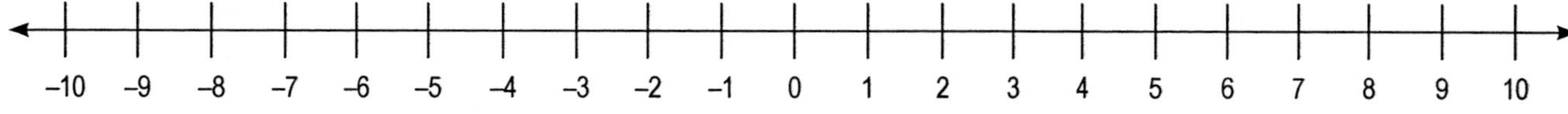

Simplify

7. $-|-11|$

8. $|-7|$

9. $-|8|$

10. A week of daily reports listing the height of a river in comparison to flood stage is given by the following table. Create a bar graph using the given information.

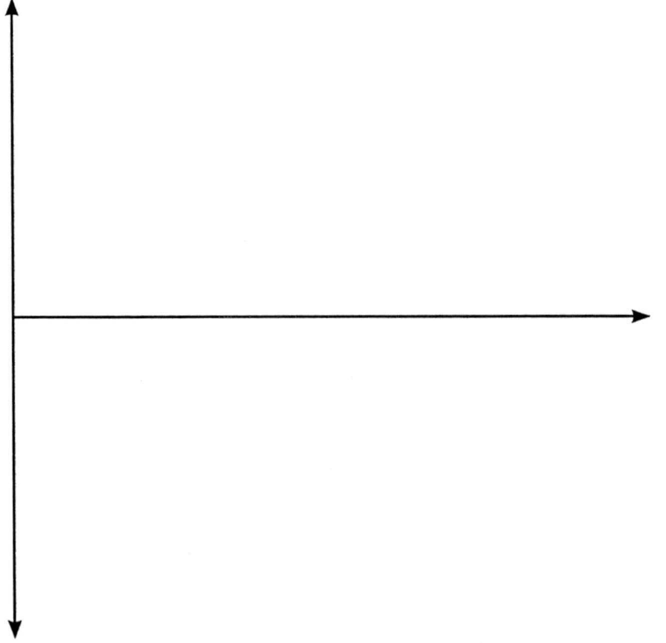

Flood stage report	
Sun.	2ft below
Mon.	3ft over
Tues.	4ft over
Wed.	2ft over
Thurs.	1ft below
Fri.	3ft below
Sat.	4ft below

2.2 ADDING INTEGERS

Objectives:

- Add integers.
- Evaluate given specific numbers.

Adding Integers

- When the numbers are both positive, the result is positive.
- When one of the numbers is negative and its absolute value is larger in distance than the second number which is positive, the result is a negative number.
- When one of the numbers is negative and its absolute value is smaller in distance than the second number which is positive, the result is positive.
- When you add two numbers which are both negative, the result is a larger negative number.

EXAMPLES:

a. $4 + (-9)$

b. $-7 + 8$

c. $-3 + (-5)$

Additive inverse of a number: when you add a number and its opposite the result is zero.

EXAMPLE:

$-4 + 4 = 0$

Extra Practice:

Find the sum.

1. $30 + (-15)$

2. $24 + (-45)$

3. $16 + (-26)$

4. $-65 + 31$

5. $4 + (-3) + (-2)$

6. $6 + (-4) + (-13) + 7$

7. $2 + (-10 + 8)$

8. $(-4 + 8) + (-11 + 4)$

9. $[-3+(-4)]+(-5+2)$

10. $[5+(-8)]+[9+(-15)]$

11. $-8+[-5+(-2)]$

12. $-675+(-456)+99$

Applications:

13. The maintenance costs, utilities, and taxes of a duplex are $900 per month. The owner of the apartments receives monthly rental payments of $450 and $380. Does this investment produce a positive cash flow each month?

14. A businessman's lunchtime workout includes jogging up ten stories of stairs in his high–rise office building. If he starts on the fourth level below ground in the underground parking garage, on what story of the building will he finish his workout?

15. After a heavy rainstorm, a river that has been 4 feet under flood stage rose 11 feet in a 48–hour period. Find the height of the river after the storm in comparison to flood stage.

16. During a battle, an army retreated 1,500 meters, regrouped, and advanced 3,500 meters. The next day, it had to retreat 1,250 meters. Find the army's distance traveled.

Review:

17. Find the area of a rectangle which has length 5 ft and width 3 ft.

18. A car with a tank that holds 15 gallons of gasoline goes 25 miles on 1 gallon. How far can it go on a full tank?

19. Find the prime factors of 4052.

HOMEWORK ASSIGNMENT: 2.2

Name: ______________________________

1. $(-9 + 12) + (-4)$

2. $(-12 + 6) + (-11 + 4)$

3. $[9 + (-12)] + (-5 + 2)$

4. $[5 + (-8)] + [8 + (-14)]$

5. $[6 + (-4)\ ] + [9 + (-15)]$

6. $-9{,}750 + (-780) + 2{,}345$

7. A student's morning workout in downtown Orlando consists of jogging up twelve stories of stairs in his high–rise condo complex. If she starts on the third level below ground in the underground parking garage, on what story of the complex will she finish her workout?

8. After a heavy rainstorm, a river that had been 6 feet under flood stage rose 13 feet in a 36–hour period. Find the height of the river after the storm in comparison to flood stage.

9. During battle, an army of foot soldiers retreated 1,000 meters, regrouped, and advanced 4,250 meters. The next day, it had to retreat 1,550 meters. Find their distance traveled.

2.3 SUBTRACTING INTEGERS

Objectives:

- Subtract integers.
- Add and subtract integers.
- Evaluate and expression by subtracting.

What is $6 - 4$? How does it compare with $6 + (-4)$?

Rules of Subtraction

If a and b are any numbers, then $a - b = a + (-b)$.

Order of Operations

When working an expression that has multiple subtraction or subtraction in combination with grouping symbols, you **MUST** apply the rules of **GEMDAS**(p. 39).

EXAMPLES:

1. $-3 - 5 - (-1)$
2. $-2 - (-6 - 5)$
3. $-1 - 5\,[5 - (-2)]$
4. $[-4 + (-8)] - (-6)$
5. $16 - 14 - (-9)$
6. $-5 - 8 - (-3)$

Extra Practice:

Find the difference.

7. $-7-6$

8. $0-(-5)$

9. $-3-(-3)$

Evaluate each expression.

10. $5-9-(-7)$

11. $6-8-(-4)$

12. $(6-4)-(1-2)$

13. $[-4+(-8)]-(-6)$

14. $[-5+(-4)]-(-2)$

Applications:

15. A diver jumps from his boat into the water and descends 50 feet. He pauses to check his equipment and then descends an additional 70 feet. Use a signed number to represent the diver's final depth.

16. Rashawn flew from his New York home to Hawaii for a week of vacation. He left blizzard conditions and a temperature of –6°, and stepped off the airplane into 85° weather. What temperature change did he experience?

17. On one lie detector test, a burglar scored –18, which indicates deception. However, on a second test, he scored –1, which is inconclusive. Find the difference in the scores.

18. Michael has $1,303 in his checking account. Can he pay his insurance premium of $676, his utility bills of $121, and his rent of $750 without having to make another deposit? Explain.

HOMEWORK ASSIGNMENT: 2.3

Name: ______________________________

1. $(6-4)-(1-7)$
2. $(-8-5)-3$
3. $-8-[4-(-9)]$
4. $[-2+(-8)]-(-6)$
5. $-1-[7-(-2)]$
6. $[-5+(-4)]-(-8)$

Applications:

7. In the state of Florida when a reading test was administered at the start of the school year, an elementary school's performance was 43 points below the county average. The principal immediately implemented a special tutorial program. At the end of the school year, retesting showed the students to be 15 points above the average. How many points did the school's reading score improve over the year?

8. Sarah flew from Orlando to visit her parents in Arkansas for a week of vacation. She left sunny conditions and a temperature of 89°, and stepped off the plane into a sudden temperature change of –4°. What was the temperature change she experienced?

9. Steven has $1,423 in his checking account. Can he pay his car insurance premium of $656, his utility bills of $233, and his rent of $725 without having to make another deposit? Explain.

10. Two of the greatest Greek mathematicians were Archimedes and Pythagoras. How many years apart were they born? How old were each of them when they died?

Review:

11. Factor 384.

12. Evaluate: $12^2 - (5 - 4)^2$

13. Solve: $3 \cdot [\ ] = 45$

14. Solve: $[\ ] - 8 = -16$

2.4 MULTIPLYING INTEGERS

Objectives:

- Multiplying Integers
- Evaluate algebraic expressions by multiplying

Rules

(+)(+) = +

(–)(–) = +

(+)(–) = (–)

(–)(+) = (–)

Multiplying by zero:

Any number multiplied by zero = zero

Powers of integers:

When a negative integer is raised to an even power, the result is positive if it is in parentheses. If you have a negative integer raised to an even power and there are no parentheses then the result is negative.

When a negative integer is raised to an odd power, the result is negative, grouping symbols or no grouping symbols.

EXAMPLES:

1. 2(–6)

2. –15 · 2

3. 0(–56)

4. –9(–7)

5. –1(–2)(–5)

6. $(-3)^4$

7. $(-4)^3$

8. -4^2

9. $(-4)^2$

Extra Practice:

10. $(-8)(-7)$

11. $-8(0)$

12. $-50 \cdot 2$

13. $-35(1)$

14. $2(3)(-5)$

15. $5(-2)(3)(-1)$

16. Find the product of –6 and the opposite of 10.

17. Find the product of the opposite of 9 and the opposite of 8.

18. $(-5)^3$

19. $(-1)^5$

20. $(-9)^2$

21. -7^2

22. -6^3

23. -11^2 and $(-11)^2$

Applications:

24. A farmer, worried about his fruit trees suffering frost damage, calls the weather service for temperature information. He is told that temperatures will be decreasing approximately 4° every hour for the next five hours. What signed number represents the total change in temperature expected over the next five hours?

25. A levee protects a town in a low–lying area from flooding. According to geologists, the banks of the levee are eroding at a rate of 2 feet per year. If something isn't done to correct the problem, what signed number indicates how much of the levee will erode during the next decade?

26. The average attendance for the WNBA Houston Comets is 8,110 a game. Suppose the team gives a sports bag, costing $3, to everyone attending the game. What signed number expresses the financial loss from this promotional giveaway?

27. A healthcare provider for a company estimates that 75 hours per week are lost by employees suffering from stress–related or preventable illness. In a 52–week year, how many hours are lost? Use a signed number to answer.

Review:

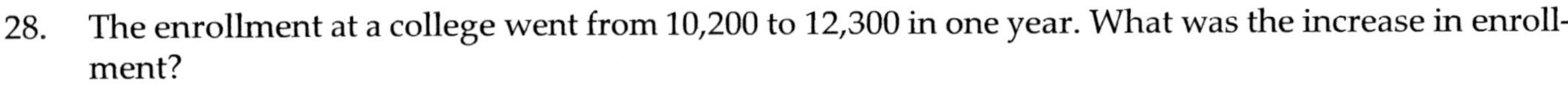

28. The enrollment at a college went from 10,200 to 12,300 in one year. What was the increase in enrollment?

29. Find the perimeter of a square with sides 6 yards long.

30. Find the area of a square with sides 6 yards long.

HOMEWORK ASSIGNMENT: 2.4

Name: ____________________

1. $-5(-3)(-4)$

2. $4(-3)(7)(-1)$

3. $-3(0)(-17)$

4. $-3(2)(-7)(-4)$

5. Find the product of –7 and the opposite of –5.

6. Find the product of the opposite of 12 and the opposite of –6.

7. $(-7)^3$

8. -9^2

9. For each of the last five years, a businesswoman in Florida has filed a $350 depreciation allowance on her income tax return, for an office computer system. What signed number represents the total amount of depreciation written off over the five–year period?

10. A healthcare provider for a company estimates that 85 hours per week are lost by employees suffering from stress–related or preventable illness. In a 52–week year, how many hours are lost? Use a signed number to answer.

2.5 DIVIDING INTEGERS

Objectives:

- Dividing Integers
- Evaluate algebraic expression by dividing

Rules for Dividing

$\frac{(+)}{(+)}=(+)$

$\frac{(-)}{(-)}=(+)$

$\frac{(+)}{(-)}=(-)$

$\frac{(-)}{(+)}=(-)$

Division with 0

1. $\frac{0}{a}=0$
2. $\frac{a}{0}=$ undefined for now.
3. $\frac{0}{0}$ is undetermined for now.

EXAMPLES:

1. $\frac{-45}{5}$
2. $\frac{60}{-20}$
3. $\frac{-21}{-3}$
4. $\frac{25}{5}$

Extra Practice:

5. $\frac{50}{-25}$

6. $\frac{-81}{9}$

7. $\frac{-45}{-15}$

8. $\frac{49}{7}$

9. Find the quotient of –45 and 9.

10. Find the quotient of –36 and –4.

Applications:

11. A mule train is to travel from a stable on the rim of the Grand Canyon to a camp on the canyon floor, approximately 5,000 feet below the rim. If the guide wants the mules to be rested after every 1,000 feet of descent, how many stops will be made on the trip?

12. The owner of a clothing store decides to reduce the price on a line of jeans that are not selling. She feels she can afford to lose $300 of projected income on these pants. By how much can she mark down each of the 20 pairs of jeans?

13. In a cost–cutting effort, a business decides to lower expenditures on salaries by $9,135,000. To do this, all of the 5,250 employees will have their salaries reduced by an equal dollar amount. How big a pay cut will each employee experience?

Review:

14. $3\left(\frac{18}{3}\right)^2 - 2(2)$

15. Find prime factorization of 210.

16. $99 = [\ \ ] - 43$

17. Sharif has scores of 55, 70, 80, and 75 on four math tests. What is his mean (average) score?

HOMEWORK ASSIGNMENT: 2.5

Name: ____________________

1. Find the quotient of –49 and 7.

2. Find the quotient of 144 and –9

3. Over a four–month period, the price of a DVR steadily fell from $400 to $340. What was the average monthly change in the price of the DVR over this period?

4. Sak's department store at the Florida Mall decided to reduce the price on a line of jeans that was not selling. The department store felt that they could afford to lose $400 of projected income on these pants. By how much can they mark down each of the 25 pairs of jeans?

5. In a cost–cutting effort, a school district decided to lower expenditures on salaries by \$8,436,000. To do this, all of the 4,216 employees will have their salaries reduced by an equal dollar amount. How big of a pay cut will each employee experience?

Review:

6. $3\left(\frac{18}{3}\right)^2 - 2(6)$

7. $78 = [\quad] - 26$

8. Kevin has test scores of 65, 76, 84, and 83 on four mathematics tests. What is his mean(average) score?

2.6 ORDER OF OPERATIONS AND ESTIMATION

Objectives:

- Simplify expressions using GEMDAS
- Evaluate an algebraic expression
- Find mean average

EXAMPLES:

1. $-5(-2)^2 - (-6)$

   $-5(4) - (-6)$

   $-20 + 6$

   -14

2. $4(2) + (-4)(-3)(-2)$

3. $-3^2 - (-3)^2$

4. $-18 + 4(-7 + 9)$

5. $-2 + (9 - 3)^2$

6. $\dfrac{-9 + 6(-4)}{(-5)^2 - 28}$

7. $-4[-2 + (5 - 9)^2]$

8. $|(-6)(5)|$

9. $|-3 + (-26)|$

10. $7 - 5\,|-1 - 6|$

Extra Practice:

11. $3^2 - 4(-2)(-1)$

12. $-3(2)^2 4$

13. $\dfrac{-7-(-3)}{2-4}$

14. $-16 - 4 \div (-2)$

15. $(7 - 5)^2 - (1 - 4)^2$

16. $(-2)^3 - (-3)(-2)$

17. $3\left(\frac{-18}{3}\right) - 2(-2)$

18. $-6 - \frac{25}{-5} + 6 \cdot 3$

19. $\frac{-3-(-7)}{2^2 - 3}$

20. $-[6 - (1 - 4)^2]$

21. $-30 + (7 - 2)^2$

22. $(5-9)^2 - (2 - 10)^2$

Make an estimate.

23. –379 + (–103) + 287

-380 - 100 +290

-480 + 290

-190

24. –3,887 + (–5, 106)

25. –36 + (–78) + 59 + (–4)

Applications:

26. In an effort to discourage her students from guessing on multiple–choice tests, a professor uses a grading scale depicted by a table. If unsure of an answer, a student does best to skip the question, because incorrect responses are penalized very heavily. Find the test score of a student who gets 12 correct and 3 wrong and leaves 5 questions blank.

Response	Value
Correct	+3
Incorrect	–4
Left blank	–1

27. The table shows the data from a chemistry experiment in spreadsheet form. To obtain a result, the chemist needs to add the values in row 1, double that sum, and then divide by the smallest value in column C. What is the final result of these calculations?

	A	B	C	D
1	12	–5	6	–2
2	15	4	5	–4
3	6	4	–2	8

HOMEWORK ASSIGNMENT: 2.6

Name: ______________________________

1. $3 + 2[10 - (1-5)]$

2. $-8 -4[1 - (12 - (-4))]$

3. $-4(3 - 4 \cdot 6)$

4. $|(-6)^2 - 3 \cdot 5|$

5. $2(5) - 7(|-2|)^2$

6. $\dfrac{3^3 - 6^2}{31 - 40}$

7. $-20 + (8 - 5)^2$

8. $(6 - 11)^2 - (1 - 4)^3$

9. In an effort to discourage her students from guessing on multiple–choice tests, a professor uses the grading scale shown in the following table. If unsure of an answer, a student does best to skip the question, because incorrect answers are penalized very heavily. Find the test score of a student who gets 14 correct, 3 wrong, and leaves 3 questions blank.

Response	Value
Correct	+4
Incorrect	–5
Left blank	–1

10. The table shows the data from a chemistry experiment in spreadsheet form. To obtain a result, the chemist needs to add the values in row 1, double that sum, and then divide that number by the smallest value in column C. What is the final result of these calculations?

	A	B	C	D
1	6	4	–2	–4
2	12	–5	6	8
3	15	4	5	–2

Review:

11. How much fencing is needed to enclose a rectangular garden with a length of 14ft and a width of 7ft?

12. What is the area of a square picture frame with sides of 15 inches?

CHAPTER 3

DECIMALS AND PERCENTS

3.1 An Introduction to Decimals

3.2 Adding and Subtracting Decimals

3.3 Multiplying Decimals

3.4 Dividing Decimals

3.5 Fractions and Decimals

3.6 Intro to Percent World Problems

3.7 Introduction to translating (is/of)

3.8 Percent Increase/Decrease

3.9 Simple Interest $I = prt$

3.1 INTRODUCTION TO DECIMALS

Objectives:

- Know the meaning of place value for a decimal number
- Write decimals in standard form
- Write decimals as fractions
- Compare decimals
- Round decimals to a given place

Place Value System for Decimals Revisited

Place Value Chart

Thousand Family			Ones Family			Decimal Family		
Hundred Thousands	Ten Thousands	Thousands	Hundreds	Tens	Ones	Tenth	Hundredth	Thousandth

Comparing Decimals

The relative sizes of a set of decimals can be determined by looking at their place values from left to right, looking for a difference in the digits.

Comparing Positive and Negative Decimals

1. Make sure both decimals have the same number of decimal places to the right of the decimal point. Write any additional zeros necessary to achieve this.
2. Compare the digits of each decimal, working from left to right.
3. When two digits differ, the decimal with the greater digit is the greater number.

Use a $<$ or $>$ symbol to compare the decimals.

EXAMPLES:

1. 113.7 _____ 113.657 2. 1.2658 _____ 1.2679

3. –703.8 _____ –703.78 4. –10.45 _____ –10.419

Rounding:

1. To round a decimal to a specified decimal place, locate the digit in that place.
2. Look to the digit to the right of the place you are rounding to.
3. Remember if the number is greater that 5, round up and drop the rest of the digits to the right; less than 5, round down and drop the rest of the digits to the right.

EXAMPLES:

Round each decimal to (a) the nearest tenth and (b) the nearest hundredth.

1. –645.132 2. 33.097 3. 9.1198

EXTRA PRACTICE:

Write using decimals:

1. Negative thirty–nine hundredths

2. Ten and fifty–six ten–thousandths

Round to the nearest tenth:

3. 0.441

4. 2.718218

Round to the nearest hundredth:

5. –808.0897

6. 33.0032

Round to the nearest thousandth:

7. 16.0995

8. 1.414213

Round the amount indicated:

9. Nearest dollar

10. Nearest ten cents

Applications:

11. The laser used in laser vision correction is so precise that each pulse can remove 39 millionths of an inch of tissue in 12 billionths of a second. Write each number as a decimal.

12. The metric system is widely used in science to measure length (meters), weight (grams), and capacity (liters). Round each decimal to the nearest hundredth.

   a. 1ft is 0.3048 meter

   b. 1mi is 1,609.344 meters

   c. 1lb is 453.59237 grams

   d. 1gal is 3.785306 liters

HOMEWORK ASSIGNMENT: 3.1

Name: ____

Write using decimals:

1. Negative twenty–seven and forty–four hundredths

2. Six and one hundred eighty–seven thousandths

Round to the nearest tenth:

3. 506.098

4. 3,987.8911

Round to the nearest hundredth:

5. –0.137

6. 64.0059

Round to the nearest thousandth:

7. 3.14159

8. 2,300.9998

Round to the amount indicated:

9. Nearest dollar

10. Nearest ten cents

Applications:

11. In July 2008, four American women set individual U.S.A. records in swimming. Their times are given below in the form *minutes: seconds*. Round each to the nearest tenth of a second.

100–meter backstroke	Natalie Coughlin	0:58.97	
200–meter breaststroke	Rebecca Soni	2:20.22	
400–meter freestyle	Katie Hoff	4:02.32	
800–meter freestyle	Katie Hoff	8:20.81	
1,500–meter freestyle	Kate Zeigler	15:42.54	

12. The metric system is widely used in science to measure length (meters), weight (grams), and capacity (liters). Round each decimal to the nearest hundredth.

a. 3 ft is 0.9144 meter

b. 4 mi is 6,437.376 meters

c. 5 lb is 2,267.96185 grams

d. 3 gal is 11.355918 liters

3.2 ADDING AND SUBTRACTING DECIMALS

Objectives:

- Add or subtract decimals
- Estimate when adding or subtracting decimals
- Simplify expressions containing decimals

Adding and Subtracting Decimals

1. Line up the decimals in a vertical stack.
2. Add or subtract the numbers as you would whole numbers.
3. Write the decimal in the result directly below the decimal points in the problem.

Adding and Subtracting Signed Decimals

- With like signs: add their absolute values and attach their common sign to the sum.
- With unlike signs: subtract their absolute values (the smaller from the larger) and attach the sign of the number with the larger absolute value to the answer.

EXAMPLES:

1. $1.903 + 0.6 + 8 + 0.78$

2. $0.07 + 35 + 0.888 + 4.1$

3. $382.8 - 277.1$

4. $30.1 - 27.122$

5. $-6.1 + (-4.7)$

6. $-21.4 - 16.75$

7. $-2.56 - (-4.4)$

8. $-4.9 - (-1.2 + 5.6)$

Applications:

1. In the sports pages of any newspaper, decimal numbers are used often.

   a. "German bobsledders set a world record today with a final run of 53.03 finishing ahead of the Italian team by only fourteen thousandths of a second." What was the time for the Italian bobsled team?

   b. "The women's figure skating title was decided by only thirty–three hundredths of a point." If the winner's point total was 102.71, what was the second–place finisher's total?

2. Forces such as water current or wind can increase or decrease the speed of an object in motion. Find the speed of each object.

   a. An airplane's speed in still air is 450mph, and it has a tail wind of 35.5 mph helping it along.

   b. A man can paddle a canoe at 5mph in still water, but he is going upstream. The speed of the current against him is 1.5mph.

HOMEWORK ASSIGNMENT: 3.2

Name: ______________________________

1. $45 + 9.9 + 0.12 + 3.02$

2. $505.01 + 23 + 0.989 + 12.07$

3. $12.98 - 3.45$

4. $30 - 11.98$

5. $-45.6 + 34.7$

6. $-7.8 + (-6.5)$

7. $(-7.2+6.3) - (-3.1 - 4)$

8. $143.3 - (-64.01)$

9. $|-14.1 + 6.9| + 8$

10. $2.3 + [2.4 - (2.5 - 2.6)]$

11. The late Florence Griffith–Joyner of the United States still holds the world record in the 100–meter-sprint: 10.49 seconds. Libby Trickett of Australia holds the world record in the 100–meter freestyle swim. Libby's time is 52.88. How much faster was Griffith–Joyner's 100 meter sprint than Trickett's 100–meter freestyle swim?

12. Complete the table by filling in the retail price of each appliance, given its cost to the dealer and the store markup.

Item	Cost	Markup	Retail Price
Refrigerator	$510.80	$105.00	
WashingMachine	$189.50	$55.50	
Dryer	$163.99	$ x	

3.3 MULTIPLYING DECIMALS

Objective:

- Multiply decimals
- Estimate when multiplying decimals
- Multiply decimals by powers of 10

1. Multiply the decimals as if they were whole numbers.
2. Find the total number of decimal places in both factors.
3. Place the decimal point in the result so that the answer has the same number of decimal places as the total found in step 2.
   - Multiplying by a power of 10, 100, 1000, etc:
   - Move the decimal to the right for each zero in 10, 100, 1000, etc.

EXAMPLES:

1. 2.74 × 4.3
2. 1.3 × 0.005
3. 2.81 × 10
4. 2.81 × 100
5. 2.81 × 1000
6. 6.6(–5.5)
7. (–1.3)2
8. –2|–4.4 + 5.6| + (–.8)2

Applications:

1. After a rainstorm, the saturated ground under a hilltop house began to give way. A survey team noted that the house dropped 0.57 inch initially. In the next two weeks, the house fell 0.09 inch per week. How far did the house fall during this three–week period?

2. In May, the water level of a reservoir reached its highest mark for the year. During the summer months, as water usage increased, it decreased 4.3 ft each month. In August, because of high temperatures, it decreased another 8.7 ft. By September, how far below the year's highest mark had the water level decreased?

3. A soccer goal measures 24ft wide by 8ft high. Major league soccer officials are proposing to increase its height by 0.75 foot.

   a. What is the area of the goal opening now?

   b. What would it be if their proposal is adopted?

   c. How much area would be added?

HOMEWORK ASSIGNMENT: 3.3

Name: ____________________

1. $(-2.13)(4.05)$

2. $6.2(100)(-0.8)$

3. $(2.3)^2$

4. $(-2.5)^2$

5. $(-6.3)(3) - (1.2)^2$

6. $(-8.1 - 7.8)(0.3 + 0.7)$

7. $-3|-8.16 + 9.9|$

8. A bakery buys various types of nuts as ingredients for cookies. Complete the table by filling in the cost of each purchase.

Type of nut	Price per pound	Pounds	Cost
Almonds	$5.95	16	
Walnuts	$4.95	25	
Peanuts	$3.85	x	

9. When billing a household, a utility company charges for the number of kilowatt–hours used. A kilowatt–hour(kwh) is a standard measure of electricity. If the cost of 1 kwh is $0.14277, what is the electric bill for a household using 719 kwh in a month? Round the answer to the nearest cent.

10. Some gas companies are required to tax the number of therms used each month by the customer. What are the taxes collected on a monthly usage of 31 therms if the tax rate is $0.00566 per therm? Round the answer to the nearest cent.

3.4 DIVIDING DECIMALS

Objectives:

- Divide decimals
- Estimate when dividing decimals
- Divide decimals by a power of 10

1. Write the problem in long division form.
2. Divide as if working with whole numbers.
3. Write the decimal point in the result directly above the decimal point in the dividend. If necessary, additional zeros can be written to the right of the dividend to allow the division to proceed.
4. If dividing with a decimal divisor, move the decimal point of the divisor so that it becomes a whole number. Make sure you also move the decimal the same number of places to the right of the number you are dividing.

EXAMPLES:

1. $47 \div 10$
2. $101.44 \div 32$
3. $0.6045 \div 0.65$
4. $2.35 \div 0.7$ Round your answer to the nearest hundredth.
5. $16.74 \div 10$
6. $16.74 \div 100$
7. $\dfrac{2.7756 + 3(-0.63)}{-0.8}$

Mean: an average of a set of data

$$\text{Mean} = \frac{\text{sum of the values}}{\text{number of values}}$$

Median: the middle number of a set of data.

1. Arrange the values in increasing order.
2. If there is an odd number of values, choose the middle value.
3. If there is an even number of values, add the middle two values and divide by 2.

Mode: the value(s) that occurs the most.

Range: the difference between the largest and smallest values.

Applications:

1. A meat slicer is designed to trim 0.05–inch–thick pieces from a sausage. If the sausage is 14 inches long, how many slices will result?

2. A computer can do arithmetic computation in 0.00003 second. How many of these computations could it do in 60 seconds?

3. Production planners have found that each squeeze of the trigger of a spray bottle emits 0.015 ounce of liquid. How many squeezes would there be in an 8.5–ounce bottle?

4. Listed below are Tara Lipinski's artistic impression scores for the long program of the women's figure skating competition at the 1998 Winter Olympics. Find the mean, median, mode, and range. Round to the nearest tenth.

   Australia 5.8

   Germany 5.8

   Ukraine 5.9

   Hungry 5.8

   U.S. 5.8

   Poland 5.8

   Austria 5.9

   Russia 5.9

   France 5.9

HOMEWORK ASSIGNMENT: 3.4

Name: ______________________________

1. $(-199.5)\div(-0.19)$

2. $(-2381.6)\div(-0.26)$

3. $\dfrac{0.0092}{0.023}$

4. $\dfrac{0.416}{0.52}$

Evaluate and round to the nearest hundredth.

5. $\dfrac{-1.2-3.4}{3(1.6)}$

6. $\dfrac{(0.5)^2-(0.3)^2}{0.005+0.1}$

7. $\dfrac{6.7 - x^2 + 1.6}{x^3}$ when $x = 0.3$

8. $\dfrac{a - b}{0.5a - 0.4b}$ when $a = 3.6$ and $b = -1.5$

9. Production planners have found that each squeeze of the trigger emits 0.015 ounce of liquid. How many squeezes would there be in a 12.6–ounce bottle?

10. In December 1998, Nkem Chukwu gave birth to eight babies in Texas Children's Hospital. Find the mean, median, and the range of their birth weights.

Ebuka (girl)	24 oz
Chidi (girl)	27 oz
Echerem (girl)	28 oz
Chima (girl)	26 oz
Odera (girl)	11.2 oz
Ikem (boy)	17.5 oz
Jioke (boy)	28.5 oz
Gorom (girl)	18 oz

3.5 FRACTIONS AND DECIMALS

Objectives:

- Write fractions as decimals
- Compare fractions and decimals
- Simplify expressions containing decimals and fractions using PEMDAS

EXAMPLES:

Write as decimals:

1. $\frac{3}{16}$

2. $\frac{4}{11}$

3. $\frac{7}{24}$

4. $8\frac{19}{20}$

Evaluate:

5. In terms of fractions: $0.53-\frac{1}{6}$

6. In terms of decimals: $0.53-\frac{1}{6}$

7. $(-0.6)^2+(2.3)\left(\frac{1}{8}\right)$

Applications:

1. While doing a tune–up, a mechanic checks the gap on one of the spark plugs of a car to be sure it is firing correctly. The owner's manual states that the gap should be inch. The gauge the mechanic uses to check the gap is in decimal notation; it registers 0.025 inch. Is the spark plug gap too large or too small?

2. A geologist weighed a rock sample at the site where it was discovered and found it weigh Later, a more accurate digital scale in the laboratory gave the weight as What is the difference in the two measurements?

Review:

1. Evaluate: $-3 + 2[-3 + (2 - 7)]$

2. Simplify: $\dfrac{\frac{7}{8}}{\frac{3}{4}}$

HOMEWORK ASSIGNMENT: 3.5

Name: ____

Write in decimal form.

1. $-8\frac{2}{3}$

2. $568\frac{23}{30}$

Use < or > to make a true statement.

3. $\frac{7}{8}\ \square\ 0.895$

4. $-9.0\overline{9}\ \square\ -9\frac{1}{11}$

Write answer in terms of fractions.

5. $0.99-\frac{5}{6}$

6. $-(0.02)\ (0.04)$

Write answer in terms of decimals.

7. $\frac{3}{8}(-3.2)+(4.5)\left(-\frac{1}{9}\right)$

8. $(-0.8)\left(\frac{1}{4}\right)+\left(\frac{1}{3}\right)(0.39)$

Round the answer to the nearest hundredth.

9. $\dfrac{3\frac{1}{5}+2\frac{1}{2}}{5.69}+3\frac{1}{4}$

Evaluate.

10. $\frac{1}{3}\,pr^2 h$ *when* $p = 3.14, r = 6,$ *and* $h = 12$

11. The amount of sunlight that comes into a room depends on the area of the windows in the room. What is the area of the window shown below?

Formula for Area of a Triangle $A=\dfrac{bh}{2}$

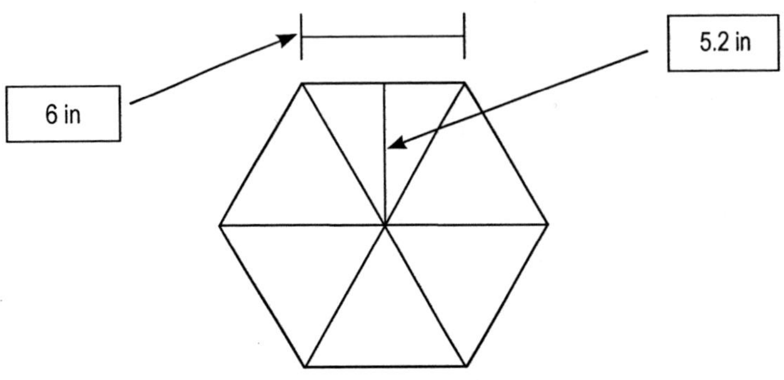

3.6 PERCENT, DECIMALS, AND FRACTIONS

Objectives:

- Understand percent
- Write percents as decimals or fractions
- Write decimals or fractions as percents
- Applications with percents, decimals, and fractions

Percent means part per one hundred.

- Change the percent to a fraction. Place the number over 100 and reduce if possible.

EXAMPLE:

1. 92%

$\frac{92}{100}=\frac{23}{25}$

2. 108%

3. 13.3%

- Change the percent to decimal. Place a decimal point at the end of the number if there is not one and move it back two places to the left.

EXAMPLE

1. 92%

.92

2. 108%

3. 13.3%

- Change decimal to percent. Multiply the number by 100 which moves the decimal point two places to the right.

1. .291

.291 * 100

29.1%

2. .06

3. 1.45

- Changing a fraction to percent. Convert the fraction portion to a decimal answer, then multiply the result by 100.

1. $\frac{5}{6}$

$\frac{5}{6}=.8\overline{3}$

.8333 * 100

83.33%

2. $2\frac{3}{4}$

3. $\frac{1}{2}$

HOMEWORK ASSIGNMENT: 3.6

Name: ______________________________

Change the percent to a fraction.

1. 94%
2. 165%
3. 62.5%

Change the percent to decimal.

4. 3.14%
5. 910%
6. 72.7%

Change decimal to percent.

7. 0.612
8. 0.0021
9. 7.03

Change the fraction to percent.

10. $\frac{21}{20}$
11. $\frac{3}{16}$
12. $\frac{3}{8}$

Applications:

1. Of the 88 keys on a piano, 36 are black.
   a. What fraction of the keys are black?
   b. What percent of the keys are black? (Round to the nearest one percent.)
2. Write as a decimal the interest rate associated with each of these accounts.
   a. Home loan: 7.75%
   b. Savings account: 5%
   c. Credit Card: 14.25%

3.7 SOLVING PERCENT PROBLEMS

Objectives:

- Write percent problems as equations

- The word "is" can be translated into an = symbol.
- The phrases, "what number" or "what percent", can be represented by a variable since there is an unknown quantity.
- The word "of" means to multiply.

Amount = *percent* · *base*

EXAMPLE:

1. What is 84% of 500?

   $x = .84(500)$

2. What is 240% of 80?

3. 38 is what percent of 40?

4. 9 is what percent of 16?

5. 6 is 75% of what number?

6. 150 is $66\frac{2}{3}$ of what number?

7. In an apartment complex, 110 of the units are currently being rented. This represents an 88% rate of occupancy. How many units are there in the complex?

HOMEWORK ASSIGNMENT: 3.7

Name: ______________________________

1. What number is 82% of 300?

2. 13 is what percent of 20?

3. What number is 36% of 250?

4. 39.6 is 44% of what number?

5. 210% of 105 is what number?

6. 9.5% of what number is 5.7?

7. After the first day of registration, 84 children had been enrolled in a new day care center. That represented 70% of the available slots. What was the maximum number of children the center could enroll?

8. On the written part of his driving test, a man answered 28 out of 40 questions correctly. If 70% correct is passing, did he pass the test?

3.8 APPLICATIONS OF PERCENT

Objectives:

- Solve applications involving percent
- Find percent increase and decrease

Taxes

EXAMPLE 1:

In the state of Texas, the sales tax is 6.25%. If an item costs \$56.35, what would the sales tax dollar amount be on that item?

EXAMPLE 2:

A waitress found that \$11.04 was deducted from her weekly gross earnings of \$240 for her federal income tax. What withholding tax rate was used?

Commissions

EXAMPLE:

An insurance sales person receives a 4.1% commission on each \$120 premium paid by a client. What is the amount of the commission on this premium?

Percent Increase/Decrease

EXAMPLE 1:

In one school district, the number of home–schooled children increased from 15 to 150 in 4 years. Find the percent of increase.

EXAMPLE 2:

One electronics store sold an MP3 player for $207.20 to a consumer, but they received it from the manufacturer for $129.50. How much of a markup did the store sell the item for?

EXAMPLE 3:

Orlando experienced a decrease in population over the ten year period from 1990 to 2000. In 1990 the population was 261,000 people. If the population decreased 10.3%, what was the population in 2000?

Discounts

EXAMPLE 1:

Sunglasses, regularly selling for $15.40, are discounted 15%. Find the sale price.

EXAMPLE 2:

An early–bird sale at a restaurant offers a $10.99 prime rib for dinner for only $7.95 if it is ordered before 6pm. Find the rate of discount.

EXAMPLE 3:

One store was running a Black Friday deal on a digital SLR camera for $288.59. The original cost is $554.99. How musch of a discount would a buyer receive if the camera is purchased?

HOMEWORK ASSIGNMENT: 3.8

Name: ______________________________

1. In the state of Arkansas, the sales tax is 4.625%. If a pair of jeans is purchased for $40 dollars, what would the sales tax dollar amount be on the item?

2. After selling a house for $198,500, a real estate agent split the 6% commission with another agent. How much did each person receive?

3. What does a ring regularly sell for if it has been discounted 20% and is on sale for $149.99?

4. What are the sale price and the discount rate for a camcorder that regularly sells for $559.97 and is being discounted $80?

5. The original cost of a CD is $14.50, but you purchased it for $18.85. What was the mark–up?

6. The cost of an MP3 player is $99.50, and with taxes your total bill was $103.48. What was the tax percentage applied to the MP3 player?

7. You purchased a computer for $3237.50, but the company paid $1850.00. How much of an increase did they charge the consumer?

8. The original price of a telescope was $99.99 and you purchased it for $86.99. How much of a discount did you receive?

9. You purchased a new SUV for $42,000.00, but with taxes applied, the final total was $43,260.00. How much tax was applied?

3.9 SIMPLE INTEREST

Objectives:

- Calculate simple interest
- Calculate compound interest

Simple Interest

Formula: I = P • r • t

I = interest

P = principal. Money invested or borrowed

r = rate (%) in decimal

t = time in years (if not in years, must convert to years)

EXAMPLE 1:

If $4200 in invested for 2 years at a rate of 4% annual interest, how much interest is earned.

EXAMPLE 2:

How much interest is paid if $3200 is borrowed at a rate of 15% for 120 days?

EXAMPLE 3:

If $375 is invested at a rate of 18% for 6 months, how much interest has it earned?

EXAMPLE 4:

You invest $950 at a rate of 12.5% for 5 years. How much interest have you earned?

EXAMPLE 5:

How much interest is paid on loan of $34,100 at 4% for 3 years?

Compound Interest

You would generally find the interest for one year and added it to the principal before proceeding to find the new interest earned on the second year. This basically becomes very tedious so you can use the compound interest formula to figure it out in one equation rather than several repeats of the simple interest formula.

$$A = P\left(1 + \frac{r}{n}\right)^{nt}$$

P = principal amount invested

r = annual interest rate expressed as a decimal

t = length of time in years

n = number of compoundings in one year

EXAMPLE 1:

As a gift for her newborn granddaughter, a grandmother opens $1000 savings account in the baby's name. The interest rate is 4.2%, compounded quarterly. Find the amount of money the child will have in the bank on her first birthday.

EXAMPLE 2:

Find the amount of interest $25,000 will earn in 10 years if it is deposited in an account at 5.99% interest, compounded daily.

EXAMPLE 3:

What is the total value of an online savings account through ING Direct when the starting balance is $7300 at 7% compounded semiannually for 3 years?

EXAMPLE 4:

You borrowed $18,000 from a loan agency at 9% compounded semiannually for 6 years. How much will you have paid at the end of the loan?

HOMEWORK ASSIGNMENT: 3.9

Name: ____________________

1. A retiree invests $5000 in a savings plan that pays 6% per year. What will be the account balance be at the end of the first year?

2. A farmer borrowed $7000 from a credit union. The money was loaned at 8.8% annual interest for 18 months. How much money did the credit union charge him for the use of the money?

3. A city is awarded a low–interest loan to help renovate the downtown business district. The $40–million loan, at 1.75%, must be repaid in 2.5 years. How much interest will the city have to pay?

4. What is the interest earned on a principal amount of $20,600 at 8% for 2 years?

5. How much interest did a retiree receive of $14,000 at a rate of 6% for 9 years?

6. What is the account balance on $1240 initial investment at 8% compounded annually for 2 years?

7. What was the total amount earned on $21,000 at 13.6% compounded quarterly for 4 years?

8. You deposited $28,600 into an account at a rate of 7.9% compounded semiannually for 2 years. What was the account balance after 2 years?

Review:

9. Divide: $-12\frac{1}{2} \div 5$.

10. Evaluate: $(0.2)^2 - (0.3)^2$

CHAPTER 4

FRACTIONS

4.1 THE FUNDAMENTAL PROPERTIES OF FRACTIONS

Objectives:

- Identify the numerator and denominator
- Write fractions to represent parts of figures or real data
- Graph fractions on a number line
- Review division properties of 0 and 1
- Write mixed numbers as improper fractions
- Write improper fractions as mixed numbers or whole numbers

Proper fractions

$\frac{1}{4}, \frac{2}{3}, \frac{98}{99}$

Improper fractions

$\frac{7}{2}, \frac{98}{97}, \frac{16}{16}, \frac{4}{1}$

- Denominator can not be zero
- Numerator and denominator can contain variables

$$\frac{x}{4}, \frac{12}{b}, \frac{x}{y}, \frac{m^2}{2mn}, \frac{2c+d}{3c^3d}$$

- Fractions can be negative

$$\frac{-a}{b} = \frac{a}{-b} = -\frac{a}{b}$$

- Not all rational numbers are integers; example, is not an integer.

Equivalent Fractions

2 fractions that represent the same number; $\frac{1}{2}, \frac{2}{4}$

Simplifying fractions:

$$\frac{a}{b} = \frac{ax}{bx}\frac{a}{b} = \frac{\frac{a}{x}}{\frac{b}{x}}$$

- Always reduce to the smallest term; factor each number and cancel the common parts.

EXAMPLES:

1. According to the calendar below, what fractional part of the month has passed? What fractional part remains?

December

$\frac{11}{31}$

~~1~~	~~2~~	~~3~~	~~4~~	~~5~~	~~6~~	~~7~~
~~8~~	~~9~~	~~10~~	~~11~~	12	13	14
15	16	17	18	19	20	21
22	23	24	25	26	27	28
29	30	31				

$\frac{20}{31}$

2. Simplify to lowest terms:

a. $\frac{60}{80}$

b. $\frac{42}{150}$

c. $\frac{45ab^2}{36ab^3}$

3. Write $\frac{2}{3}$ as an equivalent fraction with a denominator of 24.

4. Write 5 as a fraction with a denominator of 3.

5. Write $\frac{2}{5}$ as an equivalent fraction with a denominator of 30x.

EXTRA PRACTICE:

Simplify:

1. $\dfrac{20}{30}$

2. $\dfrac{24}{16}$

3. $-\dfrac{45}{54}$

4. $\dfrac{60}{108}$

5. $\dfrac{76}{28}$

6. $\dfrac{41}{51}$

7. $\dfrac{25x^2}{35x}$

8. $\dfrac{16r^2}{20r}$

9. $\dfrac{7xy}{8xy}$

10. $\dfrac{10ab}{21ab}$

11. $\dfrac{35m^3n^4}{25m^4n^3}$

12. $\dfrac{56p^4}{28p^6}$

Write as an equivalent fraction with indicated denominator.

1. $\frac{4}{5}$, denominator 35

2. $\frac{11}{16}$, denominator 32

3. $\frac{2}{7}$, denominator 14x

4. $\frac{5}{4s}$, denominator 20s

5. $\frac{2}{3}$, denominator 27t

6. $\frac{5}{12}$, denominator 36n

Write each number or algebraic expression as a fraction with the indicated denominator.

1. 4a as ninths

2. 7x as fourths

3. -10c as ninths

Applications:

1. The Earth rotates about its vertical axis once every 24 hours.

   a. What is the significance of $\frac{1}{24}$ of a rotation to us on Earth?

   b. What significance does $\frac{24}{24}$ of a revolution have?

2. On a ruler, determine how many spaces are between the numbers 0 and 1. Then determine to what number the arrow is pointing.

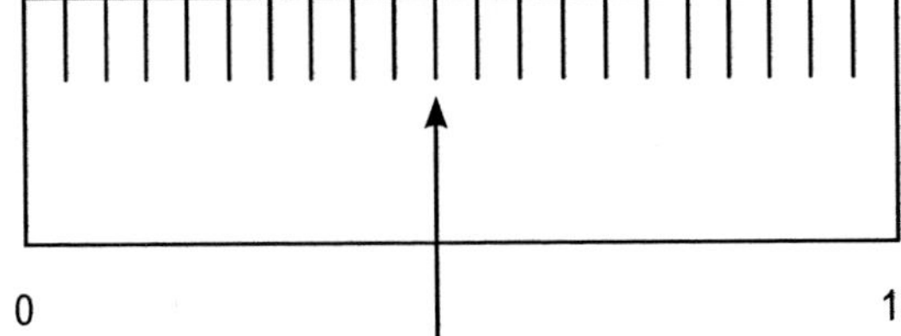

HOMEWORK ASSIGNMENT: 4.1

Name: ______________________________

1. $\frac{10rs}{8rs}$

2. $\frac{14cd}{22cd}$

3. $\frac{45f^5g^7}{25f^3g^9}$

4. $\frac{50k^5}{70k^7}$

5. $\frac{5}{4s}$, denominator 60s

6. $\frac{2}{3}$, denominator 36t

7. $\frac{5}{12}$, denominator 72n

1. The Earth rotates about its vertical axis once every 24 hours.

   a. Write a fraction that represents 8 hours has passed?

   b. What significance does of a revolution have?

2. On a ruler, determine how many spaces are between the numbers 0 and 1. Then determine to what number the arrow is pointing.

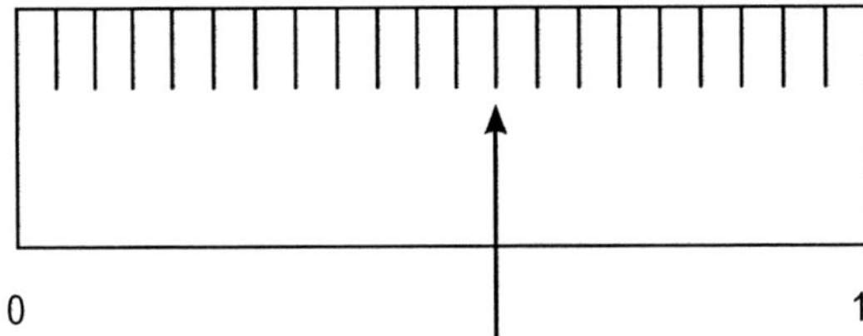

4.2 ADDING AND SUBTRACTING FRACTIONS

Objectives:

- Add or subtract like fractions
- Add or subtract unlilke fractions
- Find least common denominator of a list of fractions

Fractions with the Same Denominator

- When adding or subtracting two fractions that have the same denominator, just add or subtract the numerators and place over the common denominator based on what is given.

EXAMPLES:

1. $\frac{5}{12}+\frac{1}{12}$

2. $-\frac{9}{11}-\left(-\frac{3}{11}\right)$

3. $\frac{11}{b}-\frac{7}{b}$

4. $\frac{2}{a}+\frac{11}{a}$

Fractions with Different Denominators

- If adding the fractions: $\frac{a}{b}+\frac{a}{d}\rightarrow\frac{(a)(d)+(b)(c)}{(b)(d)}$
- If subtracting the fractions: $\frac{a}{b}-\frac{a}{d}\rightarrow\frac{(a)(d)+(b)(c)}{(b)(d)}$
- After you have finished combining the fractions, reduce if possible.
- Make sure that if the fractions have a combination of number and variable you can combine the fractions but make sure you only simplify the like terms.
- Or, you can determine what the least common denominator would be for the fractions and then multiply the numerators by the missing factor and then combine.

EXAMPLES:

1. $\frac{1}{2}+\frac{2}{5}$

2. $-6+\frac{9}{2}$

3. $\frac{y}{5}+\frac{5}{9}$

4. $\frac{6}{t}-\frac{4}{9}$

- When you have large denominators in fractions that are different, look for the LCD (least common denominator) between them; find the factors of both denominators and find what they share between each other to come up with a common denominator for both fractions.
- Need to prime factor each denominator to lowest terms, then use the product of prime factors, where each factor is used the greatest number of times in any factorization

EXAMPLES:

1. $\frac{33}{35}-\frac{11}{14}$

2. $\frac{11}{24}+\frac{7}{36}$

Comparing Fractions

- Comparing unlike fractions; convert to fractions with common denominator—preferably the LCD; then compare their numerators. The fraction with the greater numerator is the larger fraction.

EXAMPLES:

1. Which one is larger $\frac{7}{12}$ *or* $\frac{3}{5}$? (Use >3, <, or =)

2. Which one is larger $\frac{7}{8}$ *or* $\frac{31}{32}$? (Use >, <, or =)

EXTRA PRACTICE:

1. $\frac{3}{7}+\frac{1}{7}$

2. $\frac{54}{53}-\frac{52}{53}$

3. $-\frac{5}{8}-\frac{1}{3}$

4. $\frac{3}{8}-\left(-\frac{1}{6}\right)$

5. $\frac{4}{7}-\frac{1}{r}$

6. $\frac{4}{m}+\frac{2}{7}$

7. Find the difference of $\frac{11}{60}$ and $\frac{2}{45}$.

8. Find the sum of $\frac{9}{48}$ and $\frac{7}{40}$.

9. Subtract $\frac{5}{12}$ from $\frac{2}{15}$.

10. Find the sum of $\frac{11}{24}$ and $\frac{7}{36}$ increased by $\frac{5}{48}$.

Applications:

1. To ensure the effects of smog on tree development, botanists cut down a pine tree and measured the width of the growth rings for the last two years.

   a. What is the growth over this two-year period?

   b. What is the difference in the widths of the two rings?

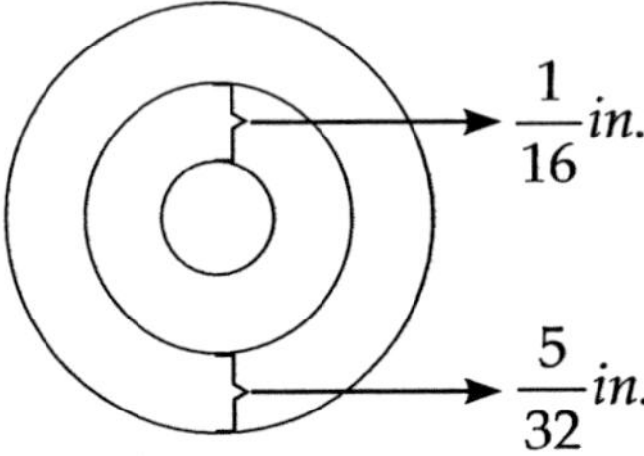

2. What is the difference in strength between a $\frac{1}{3}hp$ and a $\frac{1}{2}hp$ garage door opener?

3. A truck can safely deliver a one-ton load. Should it be used to deliver one-half ton of sand, one-third ton of gravel, and one-fifth ton of cement in one trip to a job site?

HOMEWORK ASSIGNMENT: 4.2

Name: ______________________________

1. $\frac{5}{9}+\frac{1}{9}$

2. $\frac{49}{47}-\frac{33}{47}$

3. $\frac{5}{9}-\frac{2}{s}$

4. $\frac{7}{t}+\frac{2}{5}$

5. Subtract $\frac{1}{6}$ from $\frac{2}{15}$.

6. Find the sum of $\frac{11}{27}$ from $\frac{7}{9}$ increased by $\frac{5}{54}$.

7. What is the difference in strength between a $\frac{1}{4}$ music note and a $\frac{1}{32}$ music note?

8. A truck can safely deliver a one-ton load. Should it be used to deliver one-half ton of sand, one-quarter ton of gravel, and one-fifth ton of cement in one trip to a job site?

4.3 MULTIPLYING FRACTIONS

Objectives:

- Multiply fractions
- Reduce fractions to simplest terms

- To multiply fractions, first multiply the numerators together, then multiply the denominators, then reduce if possible.

$$\frac{a}{b}\cdot\frac{c}{d}=\frac{a\cdot c}{b\cdot d}$$

EXAMPLES:

1. $\frac{5}{9}\cdot\frac{2}{3}$

$$\frac{5}{9}\cdot\frac{2}{3}=\frac{10}{27}$$

2. $\frac{5}{6}\left(-\frac{1}{3}\right)$

3. $\frac{6}{25}\cdot\frac{5}{6}$

4. $-\frac{5}{7}\cdot\left(-\frac{14}{25}\right)$

- When multiplying algebraic fractions convert everything to a fraction and then multiply.

EXAMPLES:

1. $\frac{1}{5}\cdot 5m$

$$\frac{1}{5}\cdot\frac{5m}{1}=\frac{5m}{5}=m$$

2. $\frac{5}{12y} \cdot \frac{3y}{8}$

3. $\frac{2a^3}{15b} \cdot \frac{35ab}{8a^5}$

4. $-\frac{4h^3}{5}\left(-\frac{15}{16h^3}\right)$

- **Powers of a fraction**: give the exponent for each piece inside the parentheses

EXAMPLES:

1. $\left(-\frac{3t}{4}\right)^3$

2. $\left(\frac{2m^2}{3v^3}\right)^2$

3. $\left(\frac{5rs^2}{3s^3}\right)^4$

Area of a Triangle

A= 1/2 (base)(height)

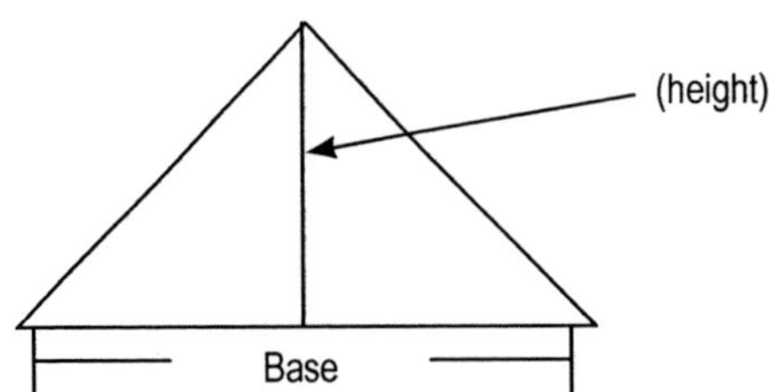

EXAMPLES:

1. What is the area of a triangle with base of 8in. and a height of 6in.?

2. What is the height of a triangle which has an area of 72 square inches and has a base of 12 inches?

EXTRA PRACTICE:

1. $\frac{5}{12} \cdot \frac{3}{4}$

2. $\left(-\frac{16}{35}\right)\left(-\frac{25}{48}\right)$

3. $\frac{1}{3} \cdot \frac{15}{16} \cdot \frac{4}{25}$

4. $6\left(-\frac{2}{3}\right)$

5. $-2\left(-\frac{7}{8}\right)$

6. $\frac{x}{2} \cdot \frac{4}{9x}$

7. $-\frac{4h^2}{5}\left(-\frac{15}{16h^3}\right)$

8. $\left(-\frac{3r}{4}\right)^3$

9. $\left(\frac{2m^2n}{3mn^2}\right)$

Find the Area of the Triangle.

10.

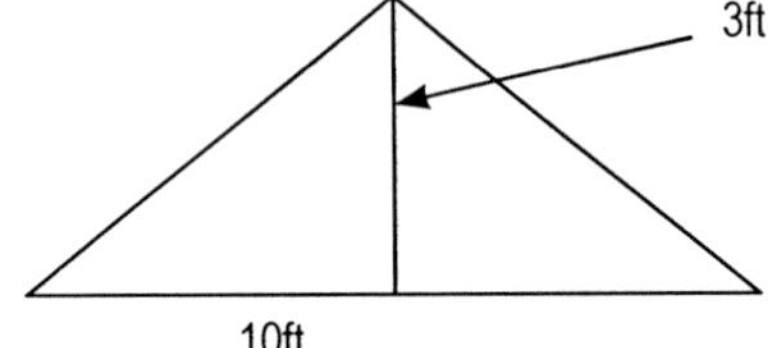

11.

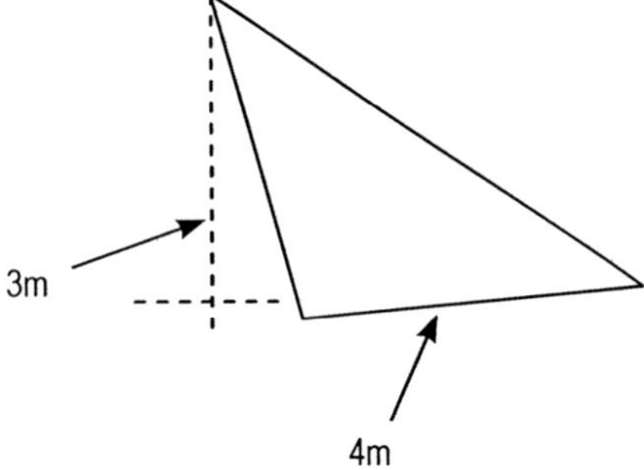

Applications:

1. Article V of the United States Constitution requires a two-thirds vote of the House of Representatives to propose a constitutional amendment. The House has 435 members. Find the number of votes needed to meet this requirement.

2. The surface of the Earth covers an area of approximately 196,800,000 square miles. About ¾ of that area is covered by water. Find the number of square miles of the surface covered by land.

HOMEWORK ASSIGNMENT: 4.3

Name: ____________________

1. $-\frac{15}{24}\cdot\frac{8}{25}$

2. $\left(\frac{3}{8}\right)\left(-\frac{2}{3}\right)\left(-\frac{12}{27}\right)$

3. $\frac{b}{12}\cdot\frac{3}{10b}$

4. $-\frac{3ef^3}{5b}\cdot\frac{10b}{e^2 f}$

5. $\left(\frac{4m}{3}\right)^2$

6. $\left(\frac{2m^2n}{3mn^2}\right)^3$

Find the Area of the Triangle.

7.

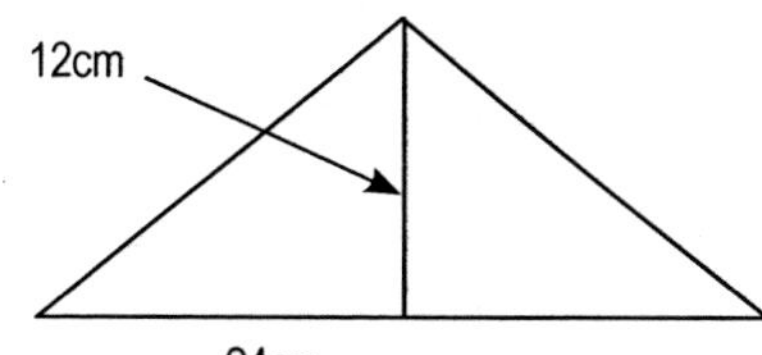

8.

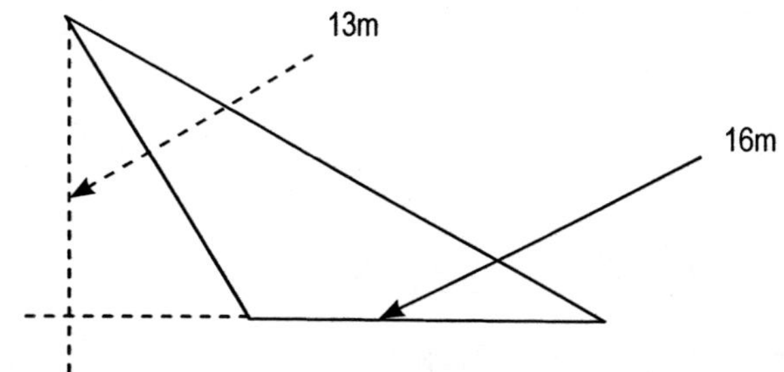

9. A design for bathroom tile is shown. Find the amount of area on the tile that is shaded.

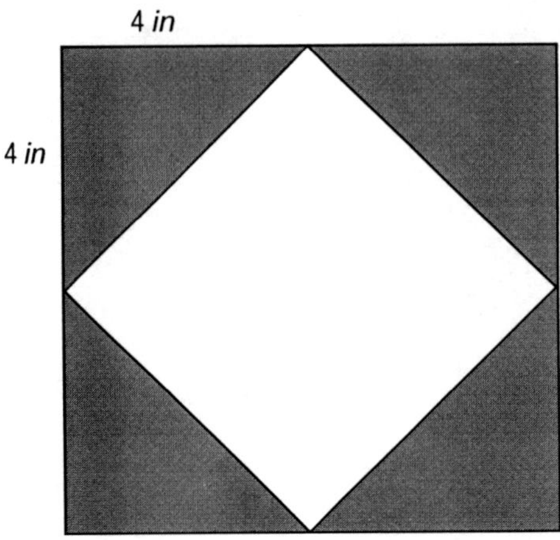

10. Estimate the area of New Hampshire, using the triangle below.

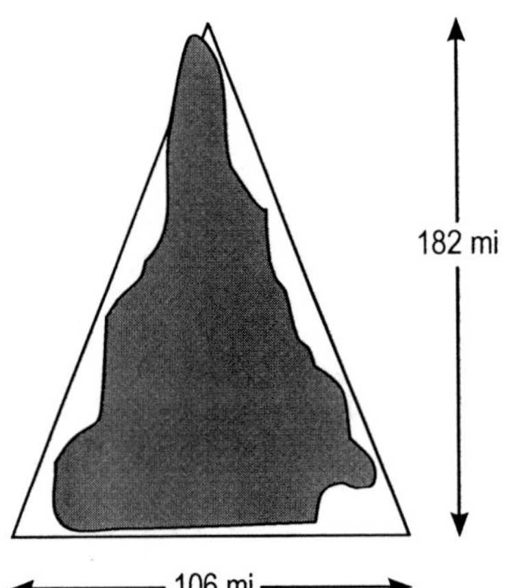

4.4 DIVIDING FRACTIONS AND ALGEBRAIC EXPRESSIONS

Objectives:

- Divide fractions

- The rule is to keep the first fraction exactly as written, change the divide sign to multiplication, and flip the second fraction by switching the numerator and denominator.

EXAMPLES:

1. $\frac{2}{3} \div \frac{7}{8}$

$$\frac{2}{3} \cdot \frac{8}{7} = \frac{16}{21}$$

2. $\frac{4}{5} \div \frac{8}{25}$

3. $\frac{2}{3} \div \left(-\frac{7}{6}\right)$

4. $-\frac{35}{16} \div (-7)$

5. $\frac{7}{4} \div \frac{3}{b}$

6. $\frac{9y^2}{10x} \div \frac{18y}{5x}$

EXTRA PRACTICE:

1. $\frac{5}{7} \div \frac{5}{6}$

2. $\frac{5}{8} \div \frac{2}{9}$

3. $\frac{7}{8} \div 120$

4. $\frac{15}{32} \div \frac{15}{32}$

5. $\frac{2x}{3} \div \frac{3}{2}$

6. $-\frac{9}{8}x \div \frac{3}{4x^2}$

7. $-12y \div \left(-\frac{3y^2}{10}\right)$

8. $-\frac{x^2}{y^2} \div \frac{x}{y}$

Applications:

1. Each lap around a stadium track is ¼ mile. How many laps would a runner have to complete to get a 26-mile workout?

2. A recipe calls for ¾ cup of flour, and the only measuring container you have holds ⅛ cup. How many ⅛ cups of flour would you need to follow the recipe?

3. A set of forestry maps divides the 6,284 acres of an old-growth forest into $^4/_5$ acre sections. How many sections do the maps contain?

4. A hardware chain purchases large amounts of nails and then packages them into $^9/_{16}$ pound bags for sale. How many of these bags of nails can be obtained from 2,871 pounds of nails?

HOMEWORK ASSIGNMENT: 4.4

Name: ______________________________

1. $\frac{7}{5} \div \frac{6}{7}$

2. $\frac{11}{16} \div \left(\frac{-9}{16}\right)$

3. $\frac{7}{8} \div 210$

4. $\frac{4a}{5} \div \frac{3}{2}$

5. $\frac{-15}{32y} \div \frac{3}{4}$

6. $-\frac{4}{3}t \div \frac{16t^2}{9}$

7. $-8x \div \left(-\frac{4x^2}{9}\right)$

8. $-\frac{k}{t^2} \div \frac{k}{t^2}$

9. Each lap around a stadium track is ¼ mile. How many laps would a runner have to complete to get a 20-mile workout?

10. A recipe calls for 1 ¼ cup of flour, and the only measuring container you have holds ⅛ cup. How many ⅛ cups of flour would you need to follow the recipe?

11. A set of forestry maps divides the 8,624 acres of an old-growth forest into $^4/_5$ acre sections. How many sections do the maps contain?

12. A hardware chain purchases large amounts of nails and packages in $^9/_{16}$ pound bags for sale. How many of these bags of nails can be obtained from 2,871 pounds of nails?

4.5 MULTIPLYING AND DIVIDING MIXED NUMBERS

Objectives:

- Multiply and Divide mixed numbers
- Perform operations on mixed numbers

- First convert the mixed fraction to an improper fraction: multiply the whole number by the denominator of the fraction portion then add the result to the numerator portion of the fraction. This becomes the numerator portion of the improper fraction. The denominator will be the original denominator of the fraction portion.

EXAMPLES:

1. Write as improper fraction: $3\frac{3}{8}$.

2. Write as a mixed number: $\frac{43}{5}$.

Multiplying and Dividing Mixed Numbers:

- First convert the mixed fractions to improper fractions and then just either multiply or divide.

EXAMPLES:

1. $9\frac{3}{5} \cdot 3\frac{3}{4}$

2. $3\frac{4}{15} \div \left(-2\frac{1}{10}\right)$

EXTRA PRACTICE:

Convert to either improper or mixed fractions.

1. $\frac{29}{5}$

2. $20\frac{4}{5}$

3. $-7\frac{1}{12}$

Multiply, divide, or evaluate:

1. $2\frac{3}{5} \cdot 1\frac{2}{3}$

2. $\left(-3\frac{1}{4}\right)\left(1\frac{1}{5}\right)$

3. $\left(1\frac{2}{3}\right)^2$

4. $3\frac{3}{4} \div 5\frac{1}{3}$

5. $\left(-\frac{5}{6}\right)\cdot(-8)\cdot\left(-2\frac{1}{10}\right)$

Applications:

1. A company advertises that its mints contain only $3\frac{1}{5}$ calories per piece. What is the calorie intake if you eat an entire package of 20 mints?

2. A cement mixer can carry $9\frac{1}{2}$ cubic yards of concrete. If it makes 8 trips to the job site, how much concrete will be delivered to the site?

3. The fire escape stairway in an office building is shown. Each riser is $7\frac{1}{2}$ inches high. If each floor is 105 inches high and the building is 43 stories tall, how many steps are in the stairway?

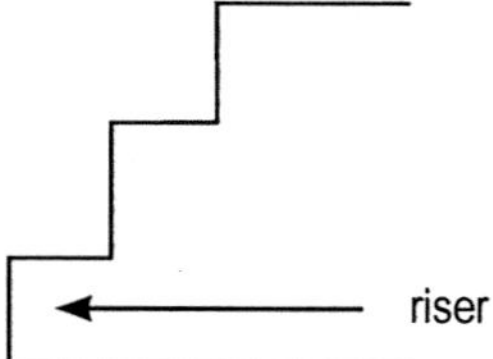

HOMEWORK ASSIGNMENT: 4.5

Name: ______________________________

Convert to either improper or mixed fractions.

1. $\frac{197}{16}$

2. $-\frac{20}{9}$

3. $90\frac{5}{6}$

Multiply or divide.

4. $-4\frac{1}{8}\left(-1\frac{7}{9}\right)$

5. $\left(3\frac{1}{2}\right)^2$

6. $\left(-1\frac{1}{5}\right)^3$

7. $-2\frac{7}{10} \div \left(-1\frac{1}{14}\right)$

8. $-20\frac{1}{4} \div \left(-1\frac{11}{16}\right)$

9. A developer donated to the county 100 of the 1,000 acres of land she owned. She divided the remaining acreage into $1\frac{1}{3}$ acre lots. How many lots were created?

10. The race tracks on which thoroughbred horses run are marked off in $\frac{1}{8}$ mile long segments called furlongs. How many furlongs are there in a $1\frac{1}{16}$ mile race?

11. How many people can be served $\frac{1}{3}$ pound hamburgers if a caterer purchases 200 pounds of ground beef?

4.6 ADDING AND SUBTRACTING MIXED NUMBERS

Objectives

- Add or subtract mixed numbers
- Perform operations on mixed numbers

- Convert the mixed numbers to improper fractions, find common denominators using LCD, then add or subtract; convert to mixed fraction if desired.
- You can also just find common denominator between the mixed fractions, add the whole numbers followed by the fractions. If an improper fraction is found convert to a mixed and add the whole piece to the larger whole number.

EXAMPLES:

1. $3\frac{2}{3}+1\frac{1}{5}$

2. $275\frac{1}{6}+81\frac{3}{5}$

3. $76\frac{11}{12}+49\frac{5}{8}$

4. $101\frac{3}{4}-79\frac{15}{16}$

5. $2300 - 129\frac{31}{32}$

EXTRA PRACTICE:

1. $8\frac{2}{7} - 3\frac{1}{7}$

2. $44\frac{3}{8} + 66\frac{1}{5}$

3. $76\frac{1}{6} - 49\frac{7}{8}$

4. $211\frac{1}{3} + 8\frac{3}{4}$

5. $7 - \frac{2}{3}$

6. $\frac{7}{3}+2$

7. $3\frac{3}{4}+5$

8. $-3\frac{3}{4}+\left(-1\frac{1}{2}\right)$

Applications:

1. A businesswomen's flight leaves Los Angeles at 8:00am and arrives in Seattle at 9:45am.
   a. Express the duration of the flight as a mixed number.

   b. Upon arrival, she boards a commuter plane at 11:15am, arriving at her final destination at 11:45am. Express the length of this flight as a fraction.

   c. Find the total time of these flights.

2. To make some draperies, Liz needs $12\frac{1}{4}$ yards of material for the den and $8\frac{1}{2}$ yards for the living room. If the material comes only in 21 yard bolts, how much will be left over after both sets of draperies are completed?

3. To repair a bad connector, Ming Lin removes $1\frac{1}{2}$ feet from the end of a 50 foot garden hose. How long is the hose after the repair?

4. A passenger ship and a cargo ship leave San Diego harbor at midnight. During the first hour, the passenger ship travels south at $16\frac{1}{2}$ miles per hour, while the cargo ship is traveling north at a rate of $5\frac{1}{5}$ miles per hour.

   a. Complete the table.

	Rate(mph) ·	Time(hr) =	Distance(mi)
Passenger ship			
Cargo ship			

   b. How far apart are the ships at 1:00am?

HOMEWORK ASSIGNMENT: 4.6

Name: ______________________________

Add or subtract.

1. $2\frac{1}{8}+3\frac{3}{8}$

2. $13\frac{5}{6}-4\frac{2}{3}$

3. $334\frac{1}{9}-13\frac{5}{6}$

4. $10\frac{1}{2}-6$

5. $12\frac{1}{2}+5\frac{3}{4}+35\frac{1}{6}$

6. $-2\frac{1}{16}+3\frac{7}{8}$

7. To make some draperies, Liz needs $10\frac{3}{4}$ yards of material for the den and $9\frac{1}{2}$ yards for the living room. If the material comes only in 21 yard bolts, how much will be left over after both sets of draperies are completed?

8. A passenger ship and a cargo ship leave San Diego harbor at midnight. During the first hour, the passenger ship travels south at $14\frac{3}{4}$ miles per hour, while the cargo ship is traveling north at a rate of $7\frac{1}{5}$ miles per hour.

   a. Complete the table.

	Rate(mph) ·	Time(hr) =	Distance(mi)
Passenger ship			
Cargo ship			

   b. How far apart are the ships at 1:00am?

4.7 ORDER OF OPERATIONS AND COMPLEX FRACTIONS

Objectives:

- Simplify complex fractions
- Review order of operations
- Evaluate expressions given values

Order of Operations

EXAMPLES:

1. $\frac{7}{8}+\frac{3}{2}\left(-\frac{1}{4}\right)^2$

2. Evaluate: $\frac{4}{3}pr^2$ *for* $p=\frac{1}{6}$ *and* $r=1\frac{1}{2}$

Complex fractions

- A fraction whose numerator or denominator, or both, contain one or more fractions or mixed numbers.

EXAMPLES:

1. $\frac{1/6}{3/8}$

2. $\dfrac{5-\dfrac{3}{4}}{1\dfrac{7}{8}}$

EXTRA PRACTICE:

1. $\dfrac{2}{3}\left(-\dfrac{1}{4}\right)+\dfrac{1}{2}$

2. $\dfrac{4}{5}-\left(-\dfrac{1}{3}\right)^2$

3. $\left(1\dfrac{2}{3}\cdot 15\right)+\left(\dfrac{7}{9}\div\dfrac{7}{81}\right)$

4. $\left|-\dfrac{3}{10}\div 2\dfrac{1}{4}\right|+\left(-2\dfrac{1}{8}\right)$

Evaluate the following using: $a=1\dfrac{3}{4}$, $b=-\dfrac{1}{5}$, $r=-1\dfrac{2}{3}$, *and* $c=-\dfrac{2}{3}$.

5. $\dfrac{1}{3}b^3-2c$

6. $3ab-4br$

Find the perimeter.

7.

$2\frac{7}{8}\,in.$

$1\frac{1}{4}\,in.$

Simplify:

8. $$\frac{-\frac{5}{6}}{-1\frac{7}{8}}$$

9. $$\frac{\frac{3}{8}+\frac{1}{4}}{\frac{3}{8}-\frac{1}{4}}$$

Evaluate for:

10. $x=-\frac{3}{4}$ and $y=\frac{7}{8}$.

11. $$\left|\frac{2x}{y-x}\right|$$

12. $$\left|\frac{y^2}{y-2}\right|$$

Applications:

1. Using a formula of $\frac{1}{2}$ ounce of sun block, $\frac{2}{3}$ ounce of moisturizing cream, and $\frac{3}{4}$ ounce of lanolin, a beautician mixes her own brand of skin cream. She packages it in $\frac{1}{4}$ ounce tubes. How many tubes can be produced using this formula?

2. At the end of a ride at an amusement park, a boat splashes into a pool of water. The time (in seconds) that it takes two pipes to refill the pool is given by $\frac{1}{\frac{1}{10}+\frac{1}{15}}$. Find the time.

HOMEWORK ASSIGNMENT: 4.7

Name: ______________________________

1. $-4\left(-\frac{1}{5}\right)-\left(\frac{1}{4}\right)\left(-\frac{1}{2}\right)$

2. $2\frac{3}{5}\left(-\frac{1}{3}\right)^2\left(\frac{1}{2}\right)$

3. $\left|\frac{2}{3}-\frac{9}{10}\right|\div\left(-\frac{1}{5}\right)$

Evaluate the following using: $a=1\frac{3}{4},\ b=-\frac{1}{5},\ r=-1\frac{2}{3}, and\ c=-\frac{2}{3}$.

4. $\frac{1}{3}b^2+c$

5. $ab-br$

Find the perimeter.

6.

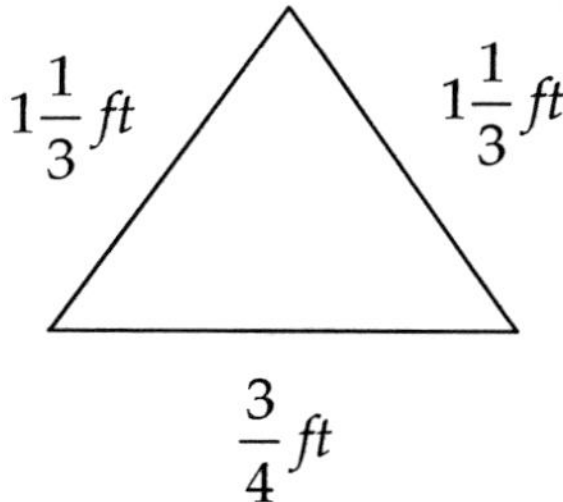

7. $$\frac{-5-\frac{1}{3}}{\frac{1}{6}+\frac{2}{3}}$$

8. $$\frac{\frac{3}{7}+\left(-\frac{1}{2}\right)}{1\frac{3}{4}}$$

Evaluate for: $x=-\frac{3}{4}$ *and* $y=\frac{7}{8}$.

9. $$\frac{x^2-y}{x+y^2}$$

10. $$\left|\frac{5x}{3y-x}\right|$$

11. Two people begin their workouts from the same point on a bike path and travel in opposite directions. One is bicycling at a rate of $7\frac{1}{5}$ mph and the other is jogging at a rate of $2\frac{1}{2}$ mph. How far apart are they in $1\frac{1}{2}$ hours? Use the table to help organize your work.

	Rate(mph)	Time(hr)	Distance(mi)
Jogger			
Cyclist			

12. A scout troop plans to hike from the campground to Glenn Peak. Since the terrain is steep, they plan to stop and rest after every $\frac{2}{3}$ mile. With this plan, how many parts will there be to this hike?

Campground → Kevin Springs = $1\frac{4}{5}$ mile

Kevin Springs → Brandon Falls = $1\frac{2}{5}$ mile

Brandon Falls → Glenn Peak = $2\frac{4}{5}$ mile

CHAPTER 5

EXPRESSIONS

5.1 VARIABLES AND ALGEBRAIC EXPRESSIONS

Objectives:

- Use properties of numbers to combine like terms
- Use distributive properties of numbers to multiply expressions
- Simplify algebraic expressions
- Find the perimeter and area of figures

Algebraic Expressions

- Variables and/or numbers that are combined with the operations of addition, subtraction, multiplication, and division to create algebraic expressions.
- Variables: are letters used to represent a missing item
- Terms: the parts of an algebraic expression that are separated by either addition or subtraction
- Coefficients: The individual number or preceding number attached to a variable

EXAMPLES:

$5(2a) = 5 \cdot 2a = 10a$

$5(2a) = 5 \cdot 2a = 10a$

$\frac{10-y}{-3} = \frac{-10}{3} + \frac{y}{3}$

$5(r-6) = (5 \cdot r) - (5 \cdot 6) = 5r - 30$

$3x + 4y = 3x + 4y$ since the variables are not the same you can not combine

$-6m^2\, n(mn) = -6m^3\, n^2$

Key Phrases Used in Translating Words into Mathematical Symbols

Addition:	Example:	Subtraction:	Example:
1. sum of p and 15	$p + 15$	1. difference of 30 an k	$30 - k$
2. 10 plus c	$10 + c$	2. 1,000 minus R	$1000 - R$
3. 5 added to a	$a + 5$	3. 15 less than w	$w - 15$
4. 4 more than r	$r + 4$	4. r decreased by 5	$r - 5$
5. 8 greater than A	$A + 8$	5. T reduced by 80	$T - 80$
6. S increased by 100	$S + 100$	6. 7 subtracted from s	$s - 7$
7. exceeds L by 20	$L + 20$	7. 2,000 less c	$2000 - c$

Multiplication:	Example:	Division:	Example:
1. product of 60 and h	$60h$	1. quotient of B and 5	$\frac{B}{5}$
2. 10 times A	$10A$	2. T divided by 50	$\frac{T}{50}$
3. twice w	$2w$	3. the ratio of h to 3	$\frac{h}{3}$
4) $\frac{1}{2}$ *of t*	$\frac{1}{2}t$	4. n split into 8 equal parts	$\frac{n}{8}$

Hint: Be careful on translating a subtraction because:

1. 5 less than $x \rightarrow x - 5$

2. x less than 5 $\rightarrow 5 - x$

Writing Algebraic Expressions to Represent Unknown Quantities

EXAMPLE 1:

A van weights p pounds. A car is 1,000 pounds lighter than the van. How many pounds does the car weight?

EXAMPLE 2:

A winning lottery ticket is worth x dollars. The payoff is to be split equally among three friends. Write an algebraic expression that represents each person's share of the prize.

EXAMPLE 3:

The sale price of a sweater is $20 less than the regular price. Choose a variable to represent the regular price. Then write an expression that represents the sale price.

EXAMPLE 4:

A plumber wants to cut a 17–foot pipe into three sections. The longest section is three times as long as the shortest sectionand the middle–sized section is 2 ft longer than the shortest section. Write an algebraic expression for each section of the pipe.

EXAMPLE 5:

On the second day of her trip, Ana drove 300 miles less than twice as far as the first day. Write an expression for the number of miles she drove each day.

EXTRA PRACTICE:

Translate:

1. The ratio of a to 100

2. The total of 5 and 12 and g

3. Two–thirds of the population p

4. 7 more than the average a

5. A man sleeps x hours per day.
   a. How many hours does he sleep in a week?
   b. In a year?

6. A rope is f feet long.
   a. Express its length in inches.
   b. Express its length in yards.

7. A couple needed to purchase 21 presents for friends and relatives on their holiday gift list. If the husband purchased g presents, how many presents did the wife need to buy?

8. A rectangle is 6 units longer than it is wide. Express the length and width of this rectangle.

9. During a sale, the regular price of a CD was reduced by $2. Express the regular price and the new sale price.

Review:

1. $-5 + (-6) + 1$

2. $-x = 4$

3. $(-5)^3$

HOMEWORK ASSIGNMENT: 5.1

Name: ______________________________

1. Translate: Seven more than four times x.

2. Translate: The quotient of nine and t is reduced by six.

3. Translate: The product of b and twelve is increased by nine.

4. Translate: Thirty–five more than the difference of 87 and g.

5. An earthquake with a reading of 8.0 on the Richter scale releases ten times as much energy (e) as an earthquake that registers 7.0 on the scale. Express the amount of energy released by an earthquake of each magnitude.

6. A rectangle is 7 units longer than it is wide. Express the length and width of the rectangle.

7. To receive a safety certificate, a crossing guard must volunteer 78 hours of service by working 6–hour shifts at a local elementary school. If she has already completed 42 hours of service, how many more 6–hour shifts must she work to receive the safety certificate?

5.2 EVALUATING ALGEBRAIC EXPRESSIONS AND FORMULAS

Objectives:

- Evaluate algebraic expressions given values for replacement
- Solve formulas for a specific variable

Evaluate Algebraic Expressions

We will be replacing variables with positive and negative numbers. It is a good idea to write parentheses around a number when it is replacing a variable in an algebraic expression.

EXAMPLE:

Evaluate each expression for $y = 5$

1. $5y - 4$ 2.

EXAMPLE:

Evaluate each expression for $t = -3$

1. $-2t + 4t^2$ 2. $-t + 2(t + 1)$ 3. $-t^2 + 16$

EXAMPLE:

Evaluate $(5rs + 4s)^2$ *for* $r = -1$ *and* $s = 5$.

Formulas Used in the Real World

Sale Price = original – discount	$s = p - d$
Retail price = cost + markup	$r = c + m$
Profit = revenue – cost	$p = r - c$
Distance = rate · time	$d = rt$
Degree Celsius = (°F–32) times five–ninths	$\frac{5}{9}(^{0}F - 32)$
Distance Fallen = 16(time)^2	$16t^2$
Mean average = sum of values divided by # of values	$A = \frac{s}{n}$

EXAMPLE 1:

The record high temperature for New Mexico is 122°F, on June 27, 1994. Convert this temperature in Celsius.

EXAMPLE 2:

Nevada's highway speed limit for trucks is 75 mph. How far would a truck travel in 2½ hours?

EXAMPLE 3:

Find the distance a rock fell in 3 seconds if it was dropped over the edge of the Grand Canyon.

EXAMPLE 4:

For the month of June, a florist's cost of doing business was $3,795. If June revenues totaled $5,115, what is her profit for the month?

EXAMPLE 5:

Ana paid \$350 for a pair of shoes. If the store's markup on the shoes was \$83, how much did the store pay for the shoes?

EXTRA PRACTICE:

Evaluate:

1. $\frac{x-8}{2}$ *for* $x = -4$

2. $a^2 + 3a - 9$ for $a = -3$

3. $-b^2 + 3b$ for $b = -4$

4. $x(5h - 1)$ for $x = -2$ and $h = 2$

5. $\frac{7v-5r}{-r}$ *for* $v=8$ *and* $r=4$

6. $-rst+2t$ for $r=-3, s=-1,$ and $t=-2$

7. Find the distance covered by a jet if it travels for 3 hours at 550mph.

8. A shopkeeper marks up the cost of every item she carries by the amount she paid for the item. If a fan costs her $27, what does she charge for the fan?

9. Find the distance a ball has fallen 2 seconds after being dropped from a tall building.

10. Find the Celsius temperature reading if the Fahrenheit reading is 113°.

11. Distance traveled:

    a. When in orbit, the space shuttle travels at a rate of approximately 17,250 miles per hour. How far does it travel in one day?

    b. The speed of light is approximately 186,000 miles per second. How far will light travel in 1 minute?

    c. The speed of a sound wave in air is about 1,100 feet per second at normal temperatures. How far does it travel in half a minute?

Review:

12. $|-2 + (-7)|$

13. $-3 - (-6)$

14. Is $t = -6$ a solution of $2t - 3 = 15$?

HOMEWORK ASSIGNMENT: 5.2

Name: ______________________________

1. $\frac{x-8}{2}$ *for* $x = -6$

2. $a^2 + 3a - 9$ for $a = -2$

3. $x(5h - 1)$ for $x = -3$ and $h = 2$

4. $\frac{7v-5r}{-r}$ *for* $v = 6$ *and* $r = 3$

5. Find the distance covered by a jet if it travels for 2.5 hours at 475mph.

6. A shopkeeper marks up the cost of every item she carries by 50% the amount she paid for the item. If a fan costs her $27, what does she charge for the fan?

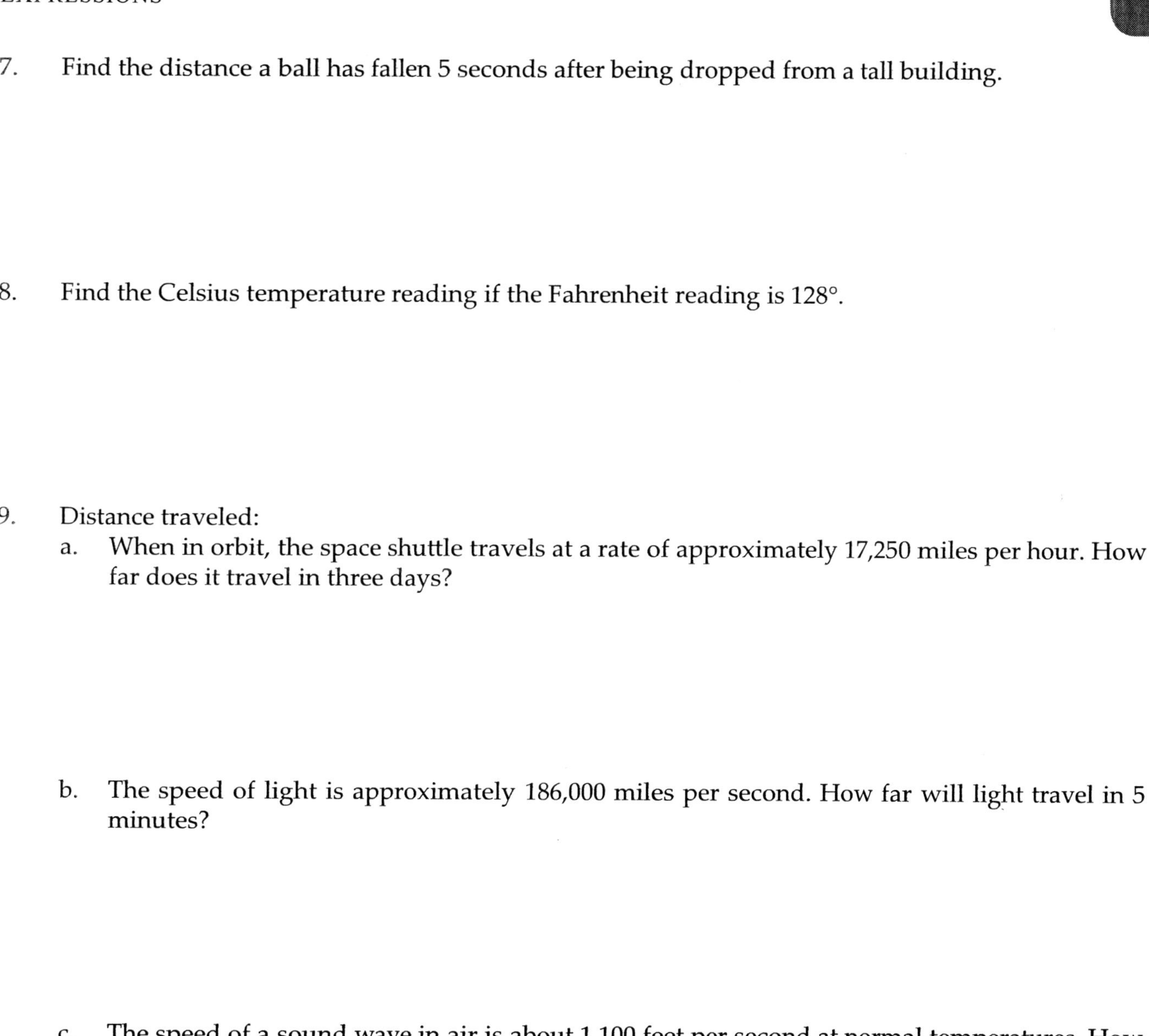

7. Find the distance a ball has fallen 5 seconds after being dropped from a tall building.

8. Find the Celsius temperature reading if the Fahrenheit reading is 128°.

9. Distance traveled:
   a. When in orbit, the space shuttle travels at a rate of approximately 17,250 miles per hour. How far does it travel in three days?

   b. The speed of light is approximately 186,000 miles per second. How far will light travel in 5 minutes?

   c. The speed of a sound wave in air is about 1,100 feet per second at normal temperatures. How far does it travel in a minute?

5.3 SIMPLIFYING ALGEBRAIC EXPRESSIONS AND FORMULAS

Objectives:

- Simplify algebraic expressions
- Distributive Property

EXAMPLE:

a. $4 \cdot 8r$ b. $-3y(-5)$ c. $2(-3d)(4a)$

Distributive Property

For any real numbers a, b and c

$a(b + c) = ab + ac$

$a(b - c) = ab - ac$

Distributing a factor of −1:

- $-(x + 8)$ does not appear to be in the proper form to use the distributive property; the number in front of the parentheses appears to be missing, but the negative sign in front of the parentheses actually represents the number (-1).

EXAMPLES:

1. $3(x + 2)$
2. $2(b - 6)$

   $(3 \cdot x) + (3 \cdot 2)$

   $3x + 6$
3. $-2(-3r - 4)$
4. $-(3 + a)$
5. $(8 + 7x + 3t)5$
6. $-(-6 - 2e)$

EXTRA PRACTICE:

Simplify:

1. $2x(3y)(3)$

2. $5r(2)(-3b)$

3. $(-1)(-2e)(-4)$

4. $3 \cdot 6j \cdot 2$

Distributive Property:

5. $4(4 - x)$

6. $-2(3e + 3)$

7. $-5(3p - 8)$

8. $-6(-1 - 3d)$

9. $3(3x - 7y + 2)$

10. $5(4 - 5r + 8s)$

11. $-3(-3z - 3x - 5y)$

12. $-(5x - 4y + 1)$

13. $-(6r - 5f + 1)$

14. $4(x + 1)$

15. $-5(7t + 2)$

16. $-8(2q - 6)$

HOMEWORK ASSIGNMENT: 5.3

Name: ______________________________

Simplify:

1. $5x(-2y)(4)$

2. $5r(7)(-2b)$

3. $-5(7p - 4)$

4. $-6(-3 - 4d)$

5. $3(-4x + 6y + 3)$

6. $-(4x - 5y + 3)$

7. $-(r - 7f + 9)$

8. $4(3x - 4z + 2)$

9. $-5(6a - 7t + 9)$

10. $-8(2q + 5s - 6)$

5.4 COMBINING LIKE TERMS

Objectives:

- Simplify expression by combining like terms
- Use the distributive property, then simplify by combing like terms

Combining like terms: to add or subtract like terms, combine their coefficients and keep the same variables with the same exponents.

Coefficient: the number in–front of a variable

Terms: the individual parts of an algebraic expression; either a number or a product of a number and a variable.

Like terms are terms with exactly the same variable raised to exactly the same powers.

EXAMPLE:

1. $5x - 2y + 7y$

2. $-8p + (-2p) + 4p$

3. $3r + 1 + p + 4$

4. $8 + 8x + x^2 - 4x$

5. $6t - 7T - 7t + 6T$

6. $-4d - (-5d)$

7. $5t - 5 - 2t + 1$

8. $-2(3x - 2) + 5x - 1$

9. $6(3y - 1) + 2(3y + 4)$

HOMEWORK ASSIGNMENT: 5.4

Name: ______________________________

Combine like terms.

1. $-5p + (-3p) + 7p$

2. $6r + 4 + p + (-2)$

3. $5 + 5x + 2x^2 - 3x$

4. $-3(2x - 1) + 3x - 4$

5. $3(6y + 1) - 2(4y - 3)$

6. $-2(2 + 3x) + 4(2x - 5)$

7. $-5(1 - 2y) - 4(6y + 5)$

8. $-3(-3y + 5) - 7(y + 6)$

9. The appropriate size of a dance floor for a given number of dancers can be determined from the table shown. Find the perimeter of each of the dance floors listed.

Slow Dancers	Fast Dancers	Size of floor (in feet)
10	7	11x11
15	10	13x13
20	15	16x16
25	18	17x17
32	22	22x22

Review:

1. Solve: $-4 \cdot [\ \ ] - 3 = -11$

2. Prime factor: 336

5.5 PROBLEM SOLVING

Objectives:

- Translate the word problems to set–up an equation but do not solve.

Addition	Subtraction	Multiplication	Division	Equals Sign
Sum	Difference	Product	Quotient	Equals
Plus	Minus	Times	Divide	Gives
Added to	Subtracted from	Multiply	Shared equally among	Is/was
More than	Less than	Twice	Per	Yields
Increased by	Decreased by	Of	Divided by	Amounts to
Total	Less	Factor	Divided into	Is equal to

Problems Involving One Unknown Quantity:

EXAMPLE 1:

In return for her services, an attorney and her client split the jury's cash award evenly. After paying her assistant $1,000, the lawyer ended up making $10,000 from the case. What was the amount of the award?

EXAMPLE 2:

Four students decided to get an apartment while attending college together. The rent for the month was $1200 plus utilities which was a fixed amount of $75 per month. How much would each individual pay if it was split equally amongst themselves?

Problems Involving Two Unknown Quantities:

EXAMPLE 3:

After receiving a donation of 400 ft of chain link fencing, the staff of a preschool decided to use it to enclose a playground that is rectangular. Find the width and the length of the playground if the length is three times the width.

Number and Value:

EXAMPLE 4:

On a night when they scored 110 points, a basketball team made only 5 free throws (worth 1 pt each). The remainder of their points came from two– and three–point baskets. If the number of baskets from the field totaled 45, how many two–point and three–point baskets did they make?

Type of basket	Number ·	Value =	Total value
3 pt			
2 pt			
Free throw			

EXTRA PRACTICE:

1. After receiving their tax refund, a husband and wife split the refunded money equally. The husband then gave $50 of his money to charity, leaving him with $70. What was the amount of the tax refund check?

2. Ten less than five times a number is the same as the number increased by six. What is the number?

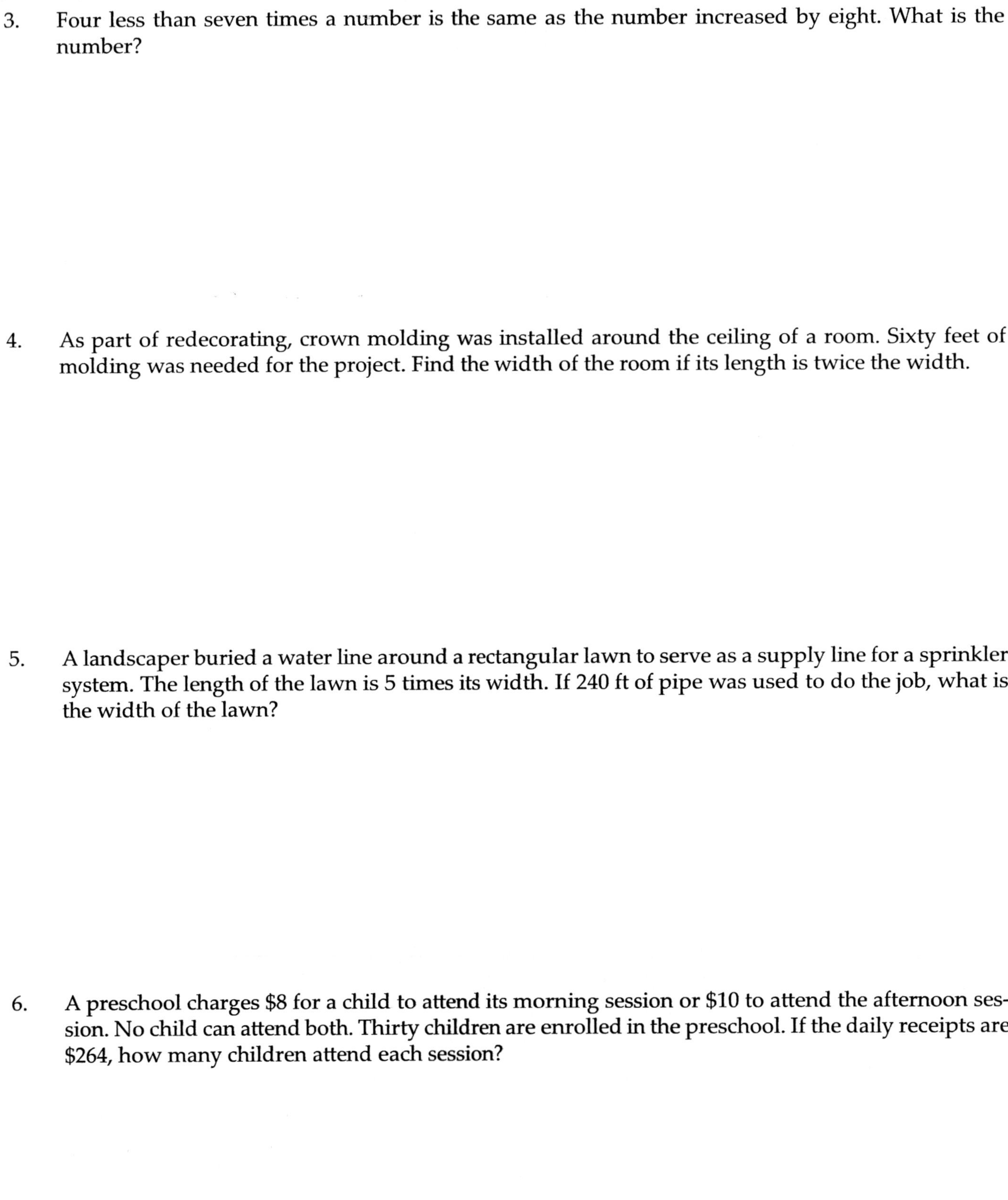

3. Four less than seven times a number is the same as the number increased by eight. What is the number?

4. As part of redecorating, crown molding was installed around the ceiling of a room. Sixty feet of molding was needed for the project. Find the width of the room if its length is twice the width.

5. A landscaper buried a water line around a rectangular lawn to serve as a supply line for a sprinkler system. The length of the lawn is 5 times its width. If 240 ft of pipe was used to do the job, what is the width of the lawn?

6. A preschool charges $8 for a child to attend its morning session or $10 to attend the afternoon session. No child can attend both. Thirty children are enrolled in the preschool. If the daily receipts are $264, how many children attend each session?

HOMEWORK ASSIGNMENT: 5.5

Name: ______________________________

Translate only. Do not Solve.

1. After receiving their tax refund, a husband and wife split the refunded money equally. The husband then gave $70 of his money to charity, leaving him with $90. What was the amount of the tax refund check?

2. Five less than ten times a number is the same as the number increased by thirteen. What is the number?

3. Seven less than four times a number is the same as the number increased by eight. What is the number?

4. As part of redecorating, crown molding was installed around the ceiling of a room. Eighty–four feet of molding was needed for the project. Find the width of the room if its length is twice the width.

5. A landscaper buried a water line around a rectangular lawn to serve as a supply line for a sprinkler system. The length of the lawn is 5 times its width. If 288 ft of pipe was used to do the job, what is the width of the lawn?

6. A preschool charges $11 for a child to attend its morning session or $13 to attend the afternoon session. No child can attend both. Thirty children are enrolled in the preschool. If the daily receipts are $364, how many children attend each session?

CHAPTER

6

EQUATIONS

6.1 Solving Equations by Addition, Subtraction, Multiplication, and Division

6.2 Solving Equations Involving Integers

6.3 Solving Equations Containing Decimals

6.4 Solving Equations Containing Fractions

6.5 Simplifying Expressions to Solve Equations

6.1 SOLVING EQUATIONS BY ADDITION , SUBTRACTION, MULTIPLICATION AND DIVISION

Objectives:

- Difference between expression and equation
- One and two step equations

Algebraic expression: Combination of terms by addition or subtraction and does not contain an equal sign

Algebraic equation: Combination of terms with only one variable to solve for and contains an equal sign

EXAMPLE:

Solve the following equations:

1. $x + 7 = 14$ subtract 7
   $-7 \ -7$
   $x = 7$

2. $x - 7 = 14$

3. $75 = b - 38$

4. $2.5 = 0.6 - b$

5. $\frac{x}{12} = 24$ Multiply by 12
   $12\left(\frac{x}{12}\right) = (24)12$
   $x = 288$

6. $13x = 39$

7. $\frac{a}{3} = 12$

8. $3.5r = 6.2$

9. $2.5 = \frac{d}{2}$

Problem Solving Strategy

- Circle any number values given.
- Box in any order of operations (add, subtract, product, etc).
- Write an equation. Then solve.
- Check your answer.

EXAMPLES:

1. The oldest artist to have a number one hit single was Louis Armstrong at age 67, with *Hello Dolly*. The youngest artist to have a number one hit single was 12-yr old Jimmy Boyd, with *I Saw Mommy Kissing Santa Claus*. What is the difference in their ages?

2. A man needs $345 for a new set of golf clubs. How much more money does he need if he now has $317?

3. The attendance at an elementary school open house was only half of what the principal had expected. If 120 people visited the school that evening, how many had she expected to attend?

EXTRA PRACTICE:

Use addition or subtraction to solve for x.

4. $23 + x = 33$

5. b - 4 = 8

6. x - 307 = 113

Applications:

7. After a student wrote a $1,500 check to pay for a car, he had a balance of $750 in his account. How much did he have in the account before he wrote the check?

Use multiplication or division to solve for x.

8. 34y = 204

9. $\frac{x}{12} = 4$

10. $400 = \frac{t}{3}$

11. $86 = 43t$

Applications:

10. Lengthy delays and skyrocketing costs caused a rapid-transit construction project to go over budget by a factor of 10. The final audit showed the project costing $540 million. What was the initial cost estimate?

11. Large sheets of commemorative stamps honoring Marilyn Monroe are to be printed. On each sheet, there are 112 stamps, with 8 stamps per row. How many rows of stamps are on a sheet?

12. A high school PE teacher had the students in her class form three-person teams for a basketball tournament. Thirty-two teams participated in the tournament. How many students were in the PE class?

HOMEWORK ASSIGNMENT: 6.1

Name:____

Add/subtract:

1. Three of Isabella's party invitations were lost in the mail, but 67 were delivered. How many invitations did she send?

2. The franchise fee and startup costs for a Taco Bell restaurant are $287,000. If an entrepreneur has $68,500 to invest, how much money will she need to borrow to open her own Taco Bell restaurant?

3. Forbes magazine estimates that in 2003, Celine Dion earned $28 million. If this was $152 million less than Oprah Winfrey's earnings, how much did Oprah earn in 2003?

Multiply/Divide:

4. $9z = 90$

5. $86 = 43t$

6. $\frac{d}{30} = 3$

7. $200 = \frac{x}{4}$

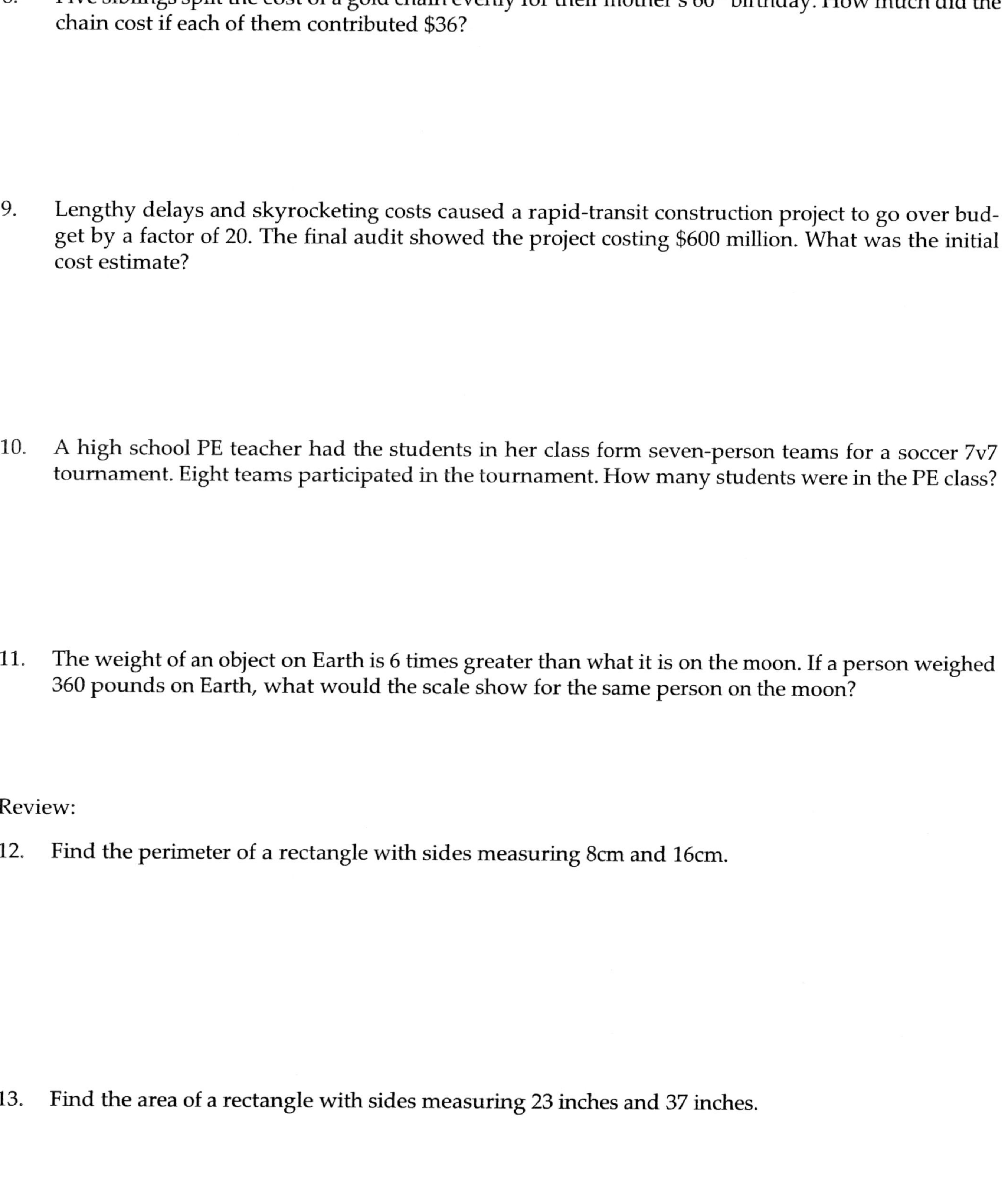

8. Five siblings split the cost of a gold chain evenly for their mother's 60th birthday. How much did the chain cost if each of them contributed $36?

9. Lengthy delays and skyrocketing costs caused a rapid-transit construction project to go over budget by a factor of 20. The final audit showed the project costing $600 million. What was the initial cost estimate?

10. A high school PE teacher had the students in her class form seven-person teams for a soccer 7v7 tournament. Eight teams participated in the tournament. How many students were in the PE class?

11. The weight of an object on Earth is 6 times greater than what it is on the moon. If a person weighed 360 pounds on Earth, what would the scale show for the same person on the moon?

Review:

12. Find the perimeter of a rectangle with sides measuring 8cm and 16cm.

13. Find the area of a rectangle with sides measuring 23 inches and 37 inches.

6.2 SOLVING EQUATIONS INVOLVING INTEGERS

Objectives:

- Use the addition/subtraction properties of equality to solve equations
- Use multiplication/division properties of equality to solve equations

EXAMPLES:

1. Solve -6 + y = -8 and check the result.

$$\begin{array}{l} -6 + y = -8 \\ +6 \quad\quad +6 \\ y = -2 \\ -6 + (-2) \_\_\_\_ -8 \\ -8 = -8 \end{array}$$

2. Solve -2 + 8 = y + 3(-4) and check the result.

3. Solve and check: -7k = 28

4. Solve and check: -40 = -8f

5. Solve and check: $\frac{t}{-3} = 4$

6. -h = -10

7. $-6b - 1 = 11$

8. $6 - 8k = -34$

9. $\frac{m}{-8} - 10 = -14$

10. $y + (-7) = -16 + 3$

11. $x - (-4) = -1 - 5$

12. $1 = \frac{m}{-5} + 6$

13. Determine if the given number is a solution of the equation $-3x - 4 = 2$; -2

14. Determine if the given number is a solution of the equation.

$$\frac{x}{-2}+5=-10;20$$

EXTRA PRACTICE:

Solve.

15. $-3v + 1 = 16$

16. $15 = -2x + (-11)$

17. $-7 - 5 - 7x = 16$

18. $8 - 2d = -5 -5$

19. $-2(5)=\frac{y}{-3}+3$

20. $-5=4+\frac{g}{-4}$

Applications:

21. After the first year of business, a manufacturer of smoke detectors found its market share 43 points behind the industry leader. Five years later, it trailed the leader by only 9 points. How many points of market share did the company pick up over the five-year span?

22. After he made a deposit of $220, a student's account was still $215 overdrawn. What was his checking account balance before the deposit?

23. During the first half of a football game, a team ran for a total of 43 yards. After a dismal second half, they ended the game with a total of -8 yards rushing. What was their rushing total in the second half?

24. A store decided that it could afford to lose some money to promote a new line of sunglasses. A $9 rebate was offered to each customer purchasing these sunglasses. If the rebate program resulted in a loss of $225 for the store, how many customers took advantage of the rebate?

25. In the first year of business, a nursery suffered a loss due to frost damage, ending the year $11,560 in the red. In the second year, it made a sizable profit. If the total profit for the first two years in business was $32,090, how much profit was made the second year?

HOMEWORK ASSIGNMENT: 6.2

Name: ______________________________

Solve.

1. $3x + 7 = 22$

2. $4 - 5x = -34$

3. $-2(6) = \frac{t}{-4} + 5$

4 $9 + 4y = 9 - 16$

5. After Shawn made a deposit of $320, his account was still $140 overdrawn. Write an algebraic equation using the given information. What was his checking account balance before the deposit?

6. Over the past year, the price of a video game player has dropped each month. If the price fell $108 this year, how much did the price drop each month on average?

7. During the holidays, a store offered a $15 gift certificate to any customer applying for its credit card. If this promotion cost the company $990, how many customers applied for credit?

8. When a group of 9 investors decided to acquire a failing bank, each had to assume an equal share of the bank's total indebtedness, which was $70,200. How much debt did each investor assume?

9. There are 12 tennis balls in one dozen, and 12 dozen in one gross. How many tennis balls are there in a shipment of 12 gross?

10. Find the perimeter and area of a rectangular garden whose length is 35 feet and width is 17 feet.

6.3 SOLVING EQUATIONS CONTAINING DECIMALS

Objectives:

- Solve equations containing decimals
- Clear the equation of any decimals.
- Use distributive property to remove parentheses if any.
- Combine like terms on either side of the equation.
- Apply the addition and subtraction properties of equality to get the variables on one side of the equation and the constant terms on the other.
- Continue to combine like terms when necessary.
- Solve for the variable.

EXAMPLES:

Solve.

1. $4.6 + t = 15.7$

$10(4.6) + 10(t) = 10(15.7)$	First eliminate decimals by multiplying by 10
$46 + 10t = 57$	Simplify
$10t = 11$	Subtract 46 from both sides
$t = 11/10$	Divide both sides by 10

2. $-1.24 = r - 0.04$

3. $\frac{y}{3} = -13.11$

4. $-22.32 = -3.1m$

5. $-4.2h + 3.14 = 1.88$

6. Simplify: $4.06a - 6.71 - 3.04a$

7. Solve: $-1.9 + 2.8x - 1.4x = 12.24$

8. $6.1b - 5.5 = 5.2b + 5.3$

9. $(4.1 + c) = -19.4$

Applications:

10. After hurricane damage estimated at $27.9 million, a county looked to three sources for relief. Local agencies contributed $6.8 million toward cleanup. A state emergency fund offered another $12.5 million. When applying for federal government help, how much should the county ask for?

11. A food dehydrator offered on a home shopping channel can be purchased by making 3 equal monthly payments. If the price is $113.25, how much is each monthly payment?

12. One 3-ounce serving of broiled ground beef has 7 grams of saturated fat. This is 14 times the amount of saturated fat in 1 cup of cooked crab meat. How many grams of saturated fat are in 1 cup of cooked crab meat?

13. The organizer of a jog-a-thon wants an archway of balloons constructed at the finish-line of the race. A company charges $100 setup fee and then 8 cents for every balloon. How many balloons will be used if $300 is spent for the decoration?

Review:

14. $-\frac{2}{3}+\frac{1}{x}$

Multiply:

15. $2\frac{1}{3}\cdot 4\frac{1}{2}$

Evaluate:

16. x^2-y^2 *where* $x=-\frac{1}{2}$ *and* $y=-1$

HOMEWORK ASSIGNMENT: 6.3

Name: ______________________________

Solve.

1. $-4.13 = r - 1.32$

2. $\frac{h}{7} = -6.23$

3. $-3.2x + 4.18 = 16.98$.

4. Solve: $-2.9 + 3.8x - 0.4x = 1.52$

5. $4(1.4+a) = -8.2$

6. A food dehydrator offered on a home shopping channel can be purchased by making 5 equal monthly payments. If the price is \$194.75, how much is each monthly payment?

7. The organizer of a jog-a-thon wants an archway of balloons constructed at the finish-line of the race. A company charges \$125 setup fee and then 5 cents for every balloon. How many balloons will be used if \$450 is spent for the decoration?

Add.

8. $-\frac{7}{8}+\frac{3}{x}$

Multiply.

9. $4\frac{2}{7}\cdot 2\frac{4}{5}$

Evaluate.

10. $3x^2-5y^2$ *where* $x=-\frac{1}{3}$ *and* $y=-2$

6.4 SOLVING EQUATIONS CONTAINING FRACTIONS

Objectives:

- Solve equations containing fractions
- Solve Equations by multiplying by the LCD
- Review adding and subtracting fractions

Using Reciprocals to Solve Equations

- In the equation $\frac{3}{4}x = 5$, the variable is multiplied by $\frac{3}{4}$. To undo this multiplication and isolate the variable, you can multiply both sides of the equation by the reciprocal of $\frac{3}{4}$ which is $\frac{4}{3}$.

EXAMPLES:

1. $-\frac{3}{2}t = 15$

2. $\frac{5}{9}t = -10$

Addition and Subtraction Properties of Equality

1. $\frac{11}{16} = a - \frac{1}{16}$

2. $y+\frac{1}{5}=\frac{2}{3}$.

Clearing Equations of Fractions

- Multiply the entire equation by a common number based from the denominators of the fractions.
- The goal is to remove all fractions and just work with whole numbers.

EXAMPLES:

1. $\frac{4}{5}p-\frac{1}{2}=\frac{3}{4}p$

2. $\frac{2}{5}x+1=\frac{1}{3}+x$

3. $\frac{y}{6}+\frac{y}{4}=-1$

EXTRA PRACTICE:

1. $\frac{7}{8}t = -28$

2. $\frac{3h}{5} = -35$

3. $x - \frac{1}{6} = \frac{2}{9}$

4. $\frac{1}{4}y - \frac{2}{3} = \frac{1}{2}$

5. $-3-2+\frac{4x}{15}=0$

Applications:

1. During a checkup, a pediatrician found that only of a child's baby teeth had emerged. The mother counted 16 teeth in the child's mouth. How many baby teeth will the child eventually have?

2. A theater usher at a Broadway musical finds that seven-eighths of the patrons attending a performance, which is 350 people, are in their seats by show time. If the show is always a complete sellout, how many seats does the theater have?

HOMEWORK ASSIGNMENT: 6.4

Name: ______________________________

Solve.

1. $\frac{2}{3}x = \frac{5}{6}$

2. $t + \frac{25}{6} = \frac{1}{8}$

3. $\frac{1}{2}x - \frac{1}{9} = \frac{1}{3}$

4. $-4 + 9 + \frac{5t}{12} = 0$

5. $6 + \frac{y}{5} = 1$

6. $\frac{2}{5}x+1=\frac{1}{3}+x$

7. $\frac{y}{6}+\frac{y}{4}=-1$

8. A telephone book consists of the white pages and the yellow pages. Two-thirds of the book consists of the white pages; the white pages number 300. Find the total number of pages in the telephone book.

9. A theater usher at a Broadway musical finds that seven-eighths of the patrons attending a performance, which is 469 people, are in their seats by show time. If the show is always a complete sellout, how many seats does the theater have?

6.5 SIMPLIFYING EXPRESSIONS TO SOLVE EQUATIONS

Objectives:

- Verify if a given number is a solution to the given equation
- Solve equations with variables on both sides of the equal sign

Checking solutions: recall that a solution of an equation is any number that satisfies the equation.

EXAMPLE:

Is -4 a solution of -4w -6 = -2 -3w?

-4(-4) -6____ -2 - 3(-4)

16 - 6 ____-2 + 12

10 = 10

Yes a solution.

EXAMPLE 1:

Is 8 a solution of -6(x +4) = -40?

Combining like terms to solve equations:

EXAMPLE 2:

8r - 6r = 16

EXAMPLE 3:

155 = d - 1 + 3d

Variables on Both Sides of an Equation

- Please make sure you isolate the variable to one side of the equation; if variables are on both sides of the equation use addition or subtraction to move one set of variables to the other side before solving the equation.

EXAMPLE 4:

9B = 3B - 18

EXAMPLE 5:

72 - 13d = 12d - 3

Removing Parentheses

- Sometimes you must do a combination of distributive property, followed by moving one set of variables to combine with its other like term before continuing to solve the equation.

EXAMPLE 6:

7(t + 5) = -70

EXAMPLE 7:

$12 + 4(x - 3) = -40$

EXAMPLE 8:

$4x = -13 - (3x + 8)$

EXAMPLE 9:

$5 - (7 - y) = -5$

EXTRA PRACTICE:

1. $3x + 6x = 54$

2. $-28 = -m + 2m$

3. $T + T - 17 = 57$

4. $403 = x - 3 + x$

5. $9t - 40 = 14t$

6. $6v + 2 = 7v - 3$

7. $-7 + 5r = 83 - 10r$

8. $2(x + 6) = 4$

9. $-16 = 2(t + 2)$

10. $-4(5t + 2) = -8$

HOMEWORK ASSIGNMENT: 6.5

Name: ______________________________

Solve.

1. $2(4y + 8) = 3(2y - 2)$

2. $3(7 - y) = 3(2y + 1)$

3. $8 + 4(x - 2) = -16$

4. $3 + 3(x - 1) = -18$

5. $-15 = 5 + 5(2x + 10)$

6. $2x + 3(x - 4) = 23$

7. $5j + 6(j + 1) = 226$

8. $10q + 3(q - 7) = 18$

9. $2q + 6(q - 4) = 24$

10. $10 - (w + 4) = -12$

11. $5 -(7 - y) = -5$

12. $10 -(x - 5) = 40$

CHAPTER 7

APPLICATIONS

7.1 Problem Solving

7.2 Simple Interest Revisited

7.1 PROBLEM SOLVING

Objectives:

- Translate the word problems to set-up an equation and then solve.

Addition	Subtraction	Multiplication	Division	Equals Sign
Sum	Difference	Product	Quotient	Equals
Plus	Minus	Times	Divide	Gives
Added to	Subtracted from	Multiply	Shared equally among	Is/was
More than	**Less than**	Twice	Per	Yields
Increased by	Decreased by	Of	Divided by	Amounts to
Total	Less	Factor	Divided into	Is equal to

Problems Involving One Unknown Quantity

EXAMPLE 1:

In return for her services, an attorney and her client split the jury's cash award evenly. After paying her assistant $1,000, the lawyer ended up making $10,000 from the case. What was the amount of the award?

EXAMPLE 2:

Four students decided to get an apartment while attending college together. The rent for the month was $1200 plus utilities which was a fixed amount of $75 per month. How much would each individual pay if it was split equally amongst themselves?

Problems Involving Two Unknown Quantities

EXAMPLE 3:

After receiving a donation of 400 ft of chain link fencing, the staff of a preschool decided to use it to enclose a playground that is rectangular. Find the width and the length of the playground if the length is three times the width.

Number and Value

EXAMPLE 4:

On a night when they scored 110 points, a basketball team made only 5 free throws (worth 1 pt each). The remainder of their points came from two- and three-point baskets. If the number of baskets from the field totaled 45, how many two-point and three-point baskets did they make?

Type of basket	Number • Value = Total value		
3 pt			
2 pt			
Free throw			

EXTRA PRACTICE:

Solve.

1. After receiving their tax refund, a husband and wife split the refunded money equally. The husband then gave \$50 of his money to charity, leaving him with \$70. What was the amount of the tax refund check?

2. Ten less than five times a number is the same as the number increased by six. What is the number?

3. Four less than seven times a number is the same as the number increased by eight. What is the number?

4. As part of redecorating, crown molding was installed around the ceiling of a room. Sixty feet of molding was needed for the project. Find the width of the room if its length is twice the width.

5. A landscaper buried a water line around a rectangular lawn to serve as a supply line for a sprinkler system. The length of the lawn is 5 times its width. If 240 ft of pipe was used to do the job, what is the width of the lawn?

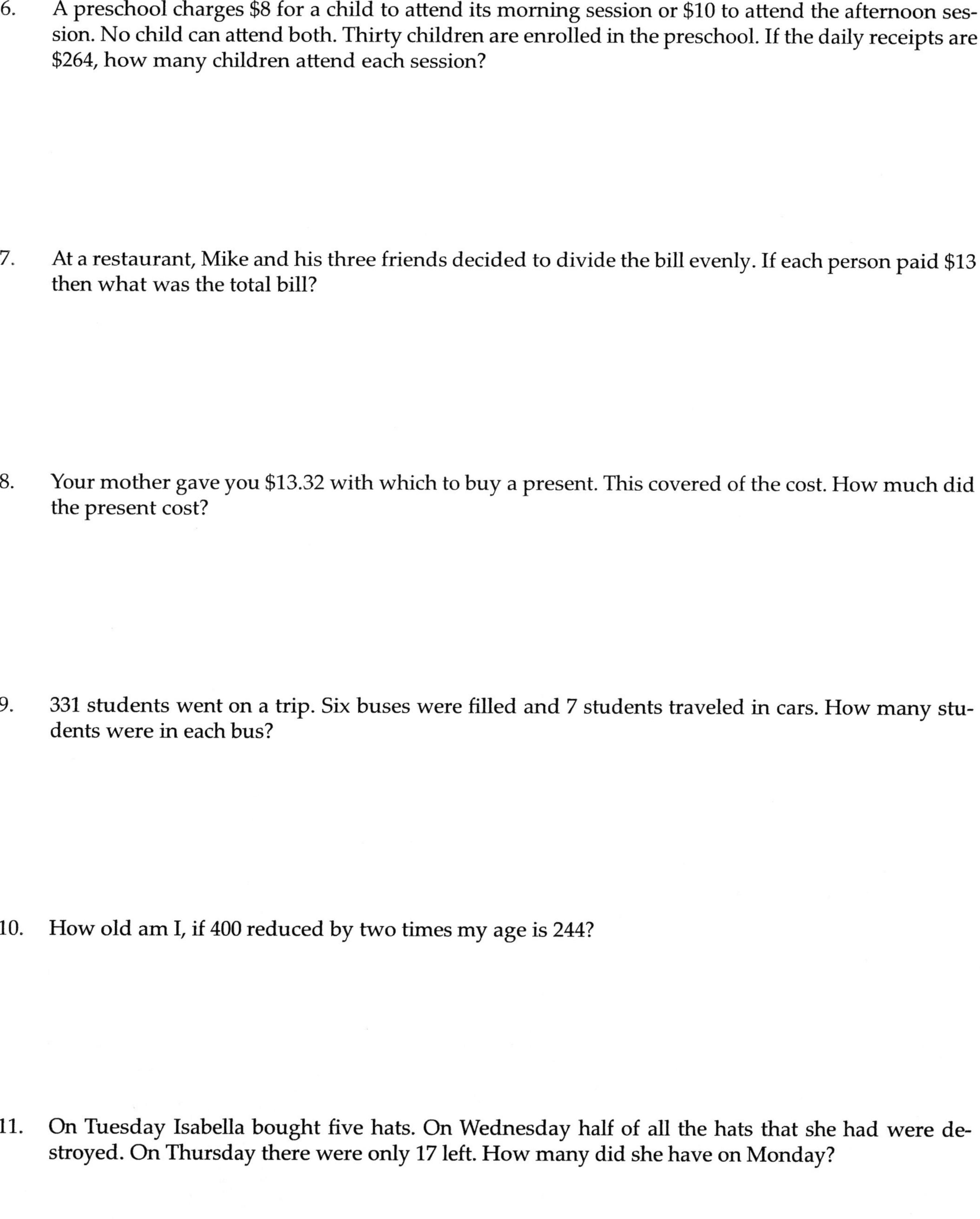

6. A preschool charges $8 for a child to attend its morning session or $10 to attend the afternoon session. No child can attend both. Thirty children are enrolled in the preschool. If the daily receipts are $264, how many children attend each session?

7. At a restaurant, Mike and his three friends decided to divide the bill evenly. If each person paid $13 then what was the total bill?

8. Your mother gave you $13.32 with which to buy a present. This covered of the cost. How much did the present cost?

9. 331 students went on a trip. Six buses were filled and 7 students traveled in cars. How many students were in each bus?

10. How old am I, if 400 reduced by two times my age is 244?

11. On Tuesday Isabella bought five hats. On Wednesday half of all the hats that she had were destroyed. On Thursday there were only 17 left. How many did she have on Monday?

HOMEWORK ASSIGNMENT: 7.1

Name: ______________________________

Solve.

1. After receiving their tax refund, a husband and wife split the refunded money equally. The husband then gave \$70 of his money to charity, leaving him with \$90. What was the amount of the tax refund check?

2. Five less than ten times a number is the same as the number increased by thirteen. What is the number?

3. Seven less than four times a number is the same as the number increased by eight. What is the number?

4. As part of redecorating, crown molding was installed around the ceiling of a room. Eighty-four feet of molding was needed for the project. Find the width of the room if its length is twice the width.

5. A landscaper buried a water line around a rectangular lawn to serve as a supply line for a sprinkler system. The length of the lawn is 5 times its width. If 288 ft of pipe was used to do the job, what is the width of the lawn?

6. A preschool charges \$11 for a child to attend its morning session or \$13 to attend the afternoon session. No child can attend both. Thirty children are enrolled in the preschool. If the daily receipts are \$364, how many children attend each session?

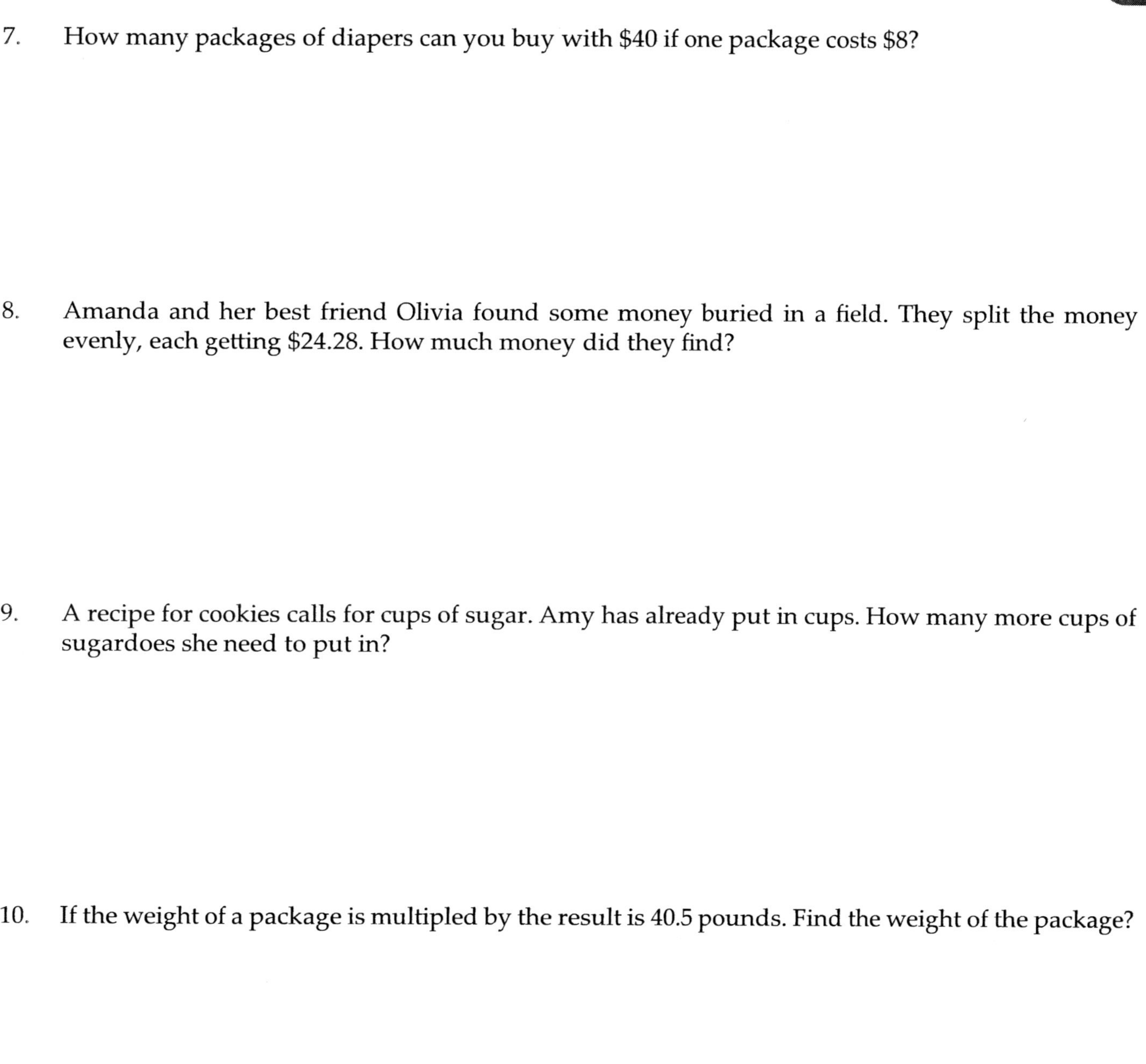

7. How many packages of diapers can you buy with $40 if one package costs $8?

8. Amanda and her best friend Olivia found some money buried in a field. They split the money evenly, each getting $24.28. How much money did they find?

9. A recipe for cookies calls for cups of sugar. Amy has already put in cups. How many more cups of sugardoes she need to put in?

10. If the weight of a package is multipled by the result is 40.5 pounds. Find the weight of the package?

11. The sum of three consecutive numbers is 72. What is the smallest of these numbers?

12. The sum of three consecutive even numbers is 48. What is the largest of the three numbers?

13. For a field trip 4 students rode in cars and the rest filled nine buses. How many students were in each bus if 472 students went on the trip?

14. The Cooking Club made some pies to sell at a soccer game to raise money for the new math books. The cafeteria contributed four pies to the sale. Each pie was then divided into five pieces and sold. There were a total of 60 pieces to sell. How many pies did the club make?

7.2 SIMPLE INTEREST REVISTED

Objectives:

- I. Calculate simple interest
- II. Calculate compound interest

Simple Interest

Formula: I = P • r • t

I = interest

P = principal. Money invested or borrowed

r = rate (%) in decimal

t = time in years (if not in years, must convert to years)

EXAMPLE 1:

If the simple interest earned was $336 for 2 years at a rate of 4% annual interest, what was the principal amount invested?

EXAMPLE 2:

What is the rate on an account that earned $1280 in a 5 year period on a principal amount of $3200?

EXAMPLE 3:

If $3375 is earned at a rate of 4.5% on an investment of $12500, how long did it take to earn that amount?

EXAMPLE 4:

You invest $950 at a rate of 12.5% for 5 years. How much interest have you earned?

Compound Interest

You would general find the interest for one year and added it to the principal before proceeding to find the new interest earned on the second year. This basically becomes very tedious so you can use the compound interest formula to figure it out in one equation rather than several repeats of the simple interest formula.

$$A = P\left(\circledcirc + \frac{r}{n}\right)^{nt}$$

P = principal amount invested

r = annual interest rate expressed as a decimal

t = length of time in years

n = number of compoundings in one year

EXAMPLE 1:

As a gift for her newborn granddaughter, a grandmother opens $1000 savings account in the baby's name. The interest rate is 4.2%, compounded quarterly. Find the amount of money the child will have in the bank on her first birthday.

EXAMPLE 2:

Find the amount of interest $25,000 will earn in 10 years if it is deposited in an account at 5.99% interest, compounded daily.

EXAMPLE 3:

What is the total value of a online savings account through ING Direct when the starting balance is $7300 at 7% comounded semiannually for 3 years?

EXAMPLE 4:

You borrowed $18,000 from a loan agency at 9% compounded semiannually for 6 years. How much will you have paid at the end of the loan?

HOMEWORK ASSIGNMENT: 7.2

Name: ______________________________

Solve.

1. A retiree invests $5000 in a savings plan that pays 6% per year. What will be the account balance be at the end of the second year?

2. A farmer borrowed $7000 from a credit union. The money was loaned at 8.8% annual interest and paid $7924back at the end of the loan. How long was the loan for?

3. A city is awarded a low-interest loan to help renovate the downtown business district. They paid back a total of $41,750,000 at the end of 2.5 years at a rate of 1.75%. How much was the original loan for?

4. What is the interest earned on a principal amount of $20,600 at 8% for 2 years?

5. How much interest did a retiree receive of $14,000 at a rate of 6% for 9 years?

6. What is the account balance on $1240 initial investment at 8% compounded annually for 2 years?

7. What was the total amount earned on $21,000 at 13.6% compounded quarterly for 4 years?

8. The total interest earned was $8806.80 at a rate of 8.2% for three years. What was the original amount invested?

9. You invested $7400 at a rate of 10.5% and earned $194.25 in interest. How long was the investment for?

CHAPTER

COORDINATE SYSTEM AND POLYNOMIALS

8.1 THE RECTANGULAR COORDINATE SYSTEM

The Rectangular Coordinate System

Quadrant : both the x– and y–coordinate points are (+)

Quadrant : the x–coordinate is (–) and the y–coordinate is (+)

Quadrant : both the x– and y–coordinate points are (–)

Quadrant : the x–coordinate is (+) and the y–coordinate is (–)

Understanding the ordered pair: , the number in the x–position will run left to right on the coordinate grid like you would normally plot on a regular number line; the number in y–position will run up and down as if riding an elevator. Remember to look at the sign value either (+) or (–) to determine where on the grid is the point located.

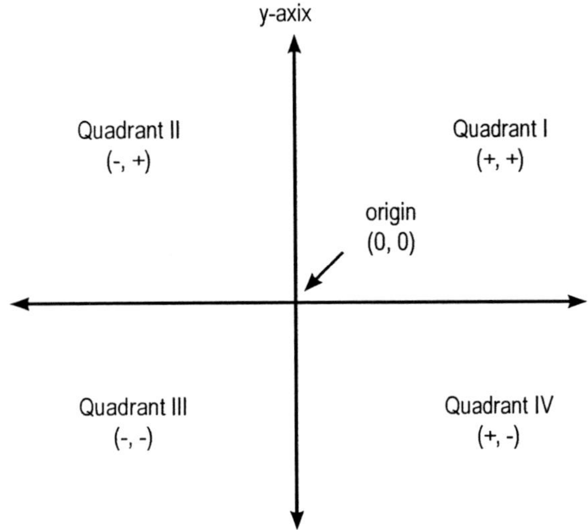

EXAMPLES:

Plot the following points (–3, 0), (0, 4), (2.5,–4), (–3.5, –2.5), (0, –2.5), (2, 2), (4, 0), (–3, 4)

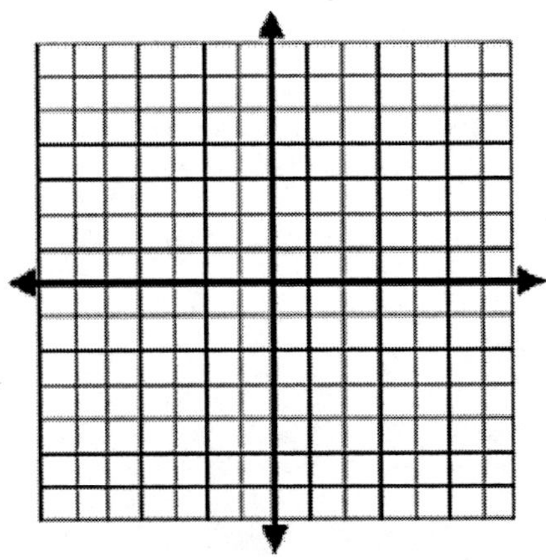

Solutions of Equations in Two Variables

We have worked with equations containing one variable. For example: . We will now be working with equations containing two variables, such as . When you are asked to see it the value is a solution to the equation, replace the variables with their numerical value to see if it balances the equation.

EXAMPLE 1:

Determine whether **(2, 4)** is a solution of: **$3x + y = 12$.**

EXAMPLE 2:

Determine whether **(0, 6)** is a solution of: **$4x + 3y = 24$.**

Completing a table:

Remember, since the table is giving you a set of x–values to use, replace the (x) variable in the equation and solve for the (y) variable.

EXAMPLES:

Complete the table of solutions for: **$5x - 4y = 20$.**

x	y	(x, y)
0		
1		
2		

Complete the table of solutions for: **$2x + 3y = 5$.**

x	y	(x, y)
0		
1		
2		

EXTRA PRACTICE:

1. If x, then $y =$ ______, in the equation: **3x + y = 12.**

2. If y, then $x =$ ______, in the equation: **3x + y = 12.**

Complete the table:

$7x - y = 21$

x	y	(x, y)
0		
2		
3		

Plot the points:

(−1.5, 2), (3, −2.5), (0, −3), (3, 0),
(0, 0), (−0.5, 2.5), (5, −5), (−5, 5)

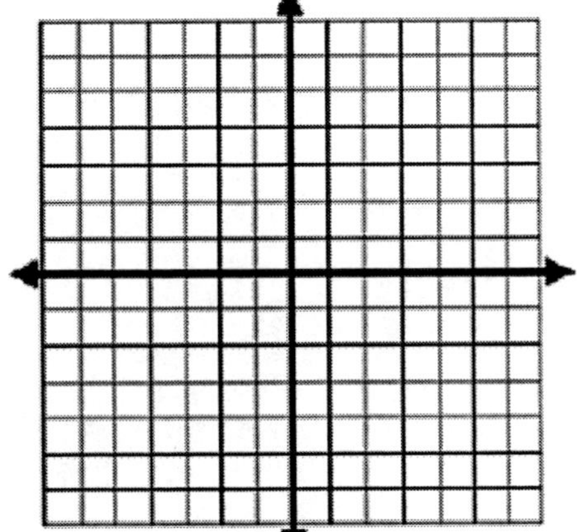

Determine the (x, y) coordinates of the following points shown on the graph.

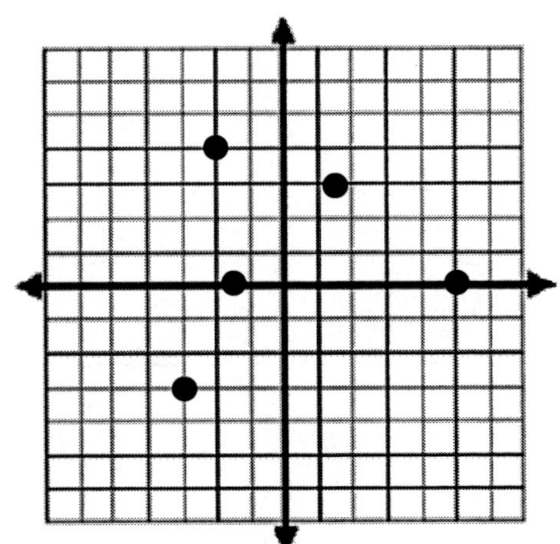

HOMEWORK ASSIGNMENT: 8.1

Name: ______________________________

1. Plot the following points (−1, 2), (1, 4), (1.5,−3), (−4.5, −2.5), (0, −3.5), (3, 3), (1, 0), (−3,3)

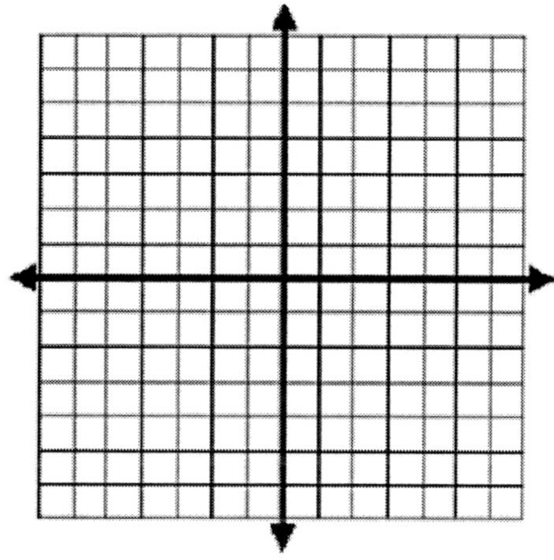

2. Complete the table of solutions for:

x	y	(x, y)
0		
1		
2		

3. Complete the table of solutions for: $4x - 3y = 12$

x	y	(x, y)
0		
3		
6		

4. Determine whether (1, 3) is a solution of: $x - 7y = 21$.

5. Determine whether (4, 0) is a solution of: $3x + 4y = 12$.

8.2 GRAPHING LINEAR EQUATIONS

Graphing Linear Equations

Here are the steps to graph a line

1. Solve for y if necessary
2. Choose any three numbers for x;
3. Replace the value of x in the equation to find y.
4. Plot the points on the graph and draw the line.

EXAMPLE 1:

$y - 3x = -2$

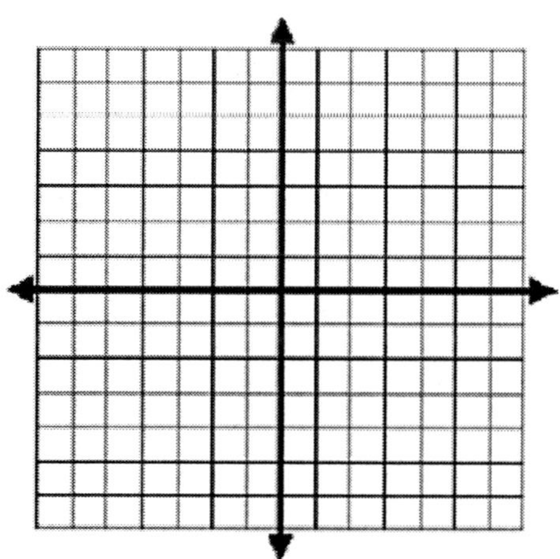

EXAMPLE 2:

$y - x = 1$

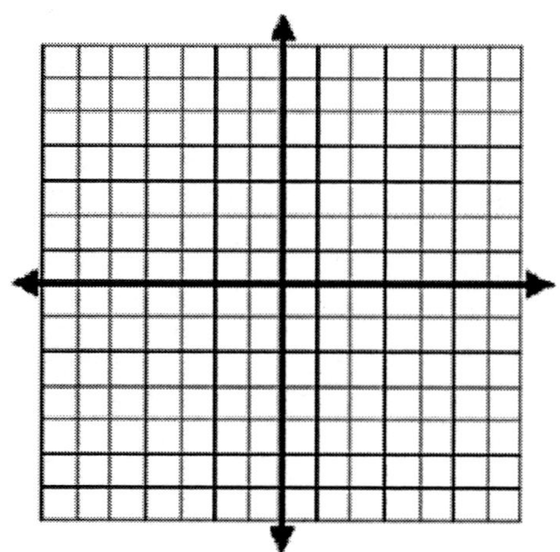

EXAMPLE 3:

$y = -3x + 5$

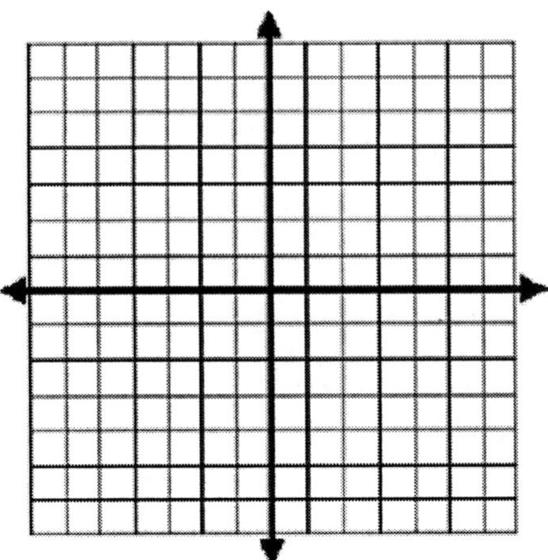

Using the Intercepts to Graph Linear Equations

x –intercept: the point where the graph crosses the x –axis (horizontal intercept); to find the x–intercept you must make the y = 0; (Hint: you can place your pointer finger over the y–variable and solve for the x–variable only)

y –intercept: the point where the graph crosses the y –axis (vertical intercept); to find the y–intercept you must make the x = 0; (Hint: you can place your pointer finger over the x–variable and solve for the y–variable only)

EXAMPLE:

Graph the line using the intercept method. $2x + 3y = 6$

x	y
0	
	0

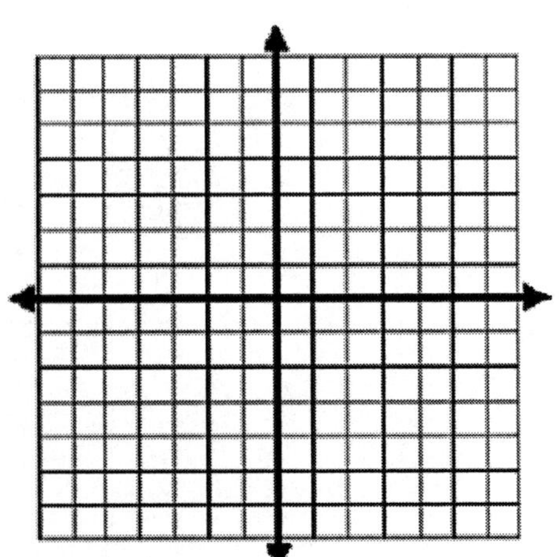

Graphing Horizontal and Vertical Lines

→ Horizontal Lines

The graph of a horizontal line is: **y = k** (k is a constant; a number)

EXAMPLE:

Graph $y = 2$

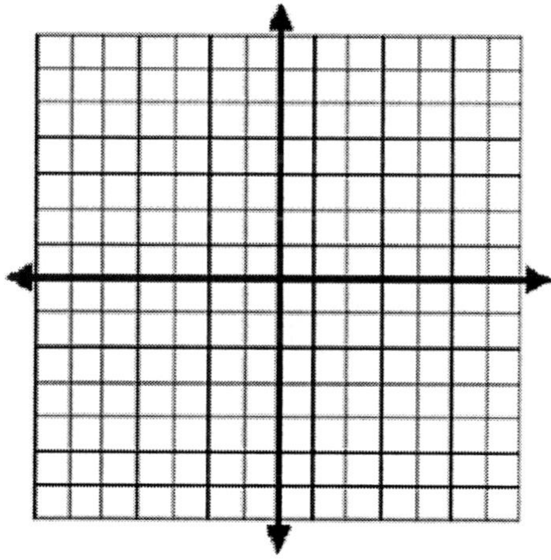

Graph $y = -6$

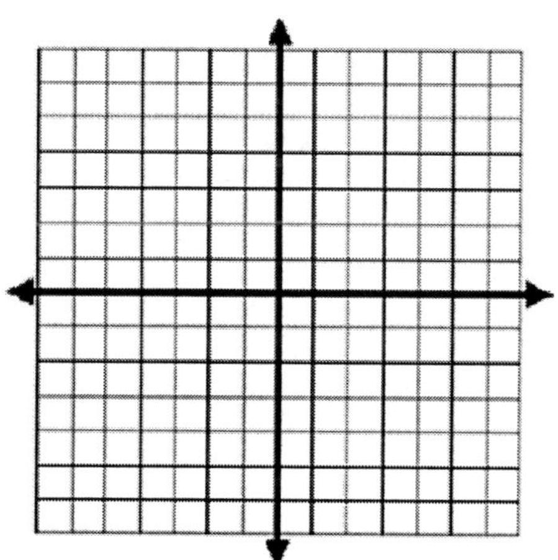

Vertical Lines

The graph of a vertical line is: **x = k** (k is a constant)

EXAMPLE:

Graph $x = 5$

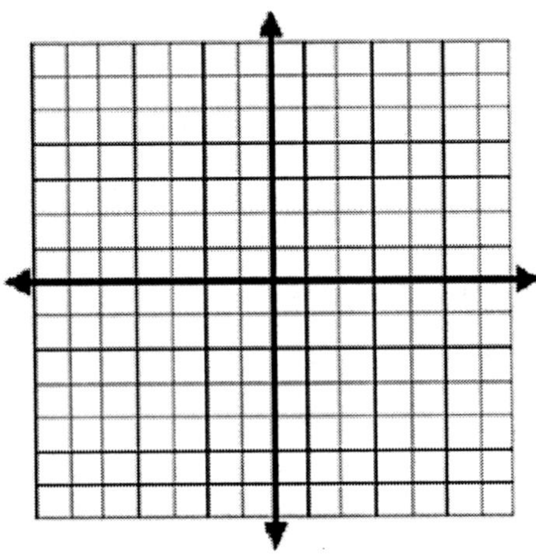

Graph $x = -7$

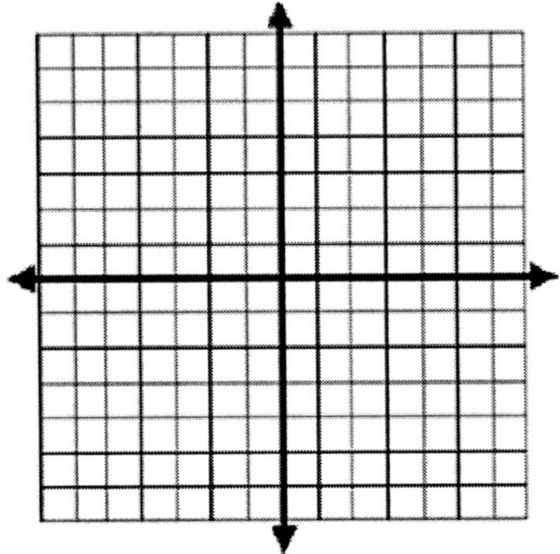

EXTRA PRACTICE:

1. Graph $3x - 5y = 15$

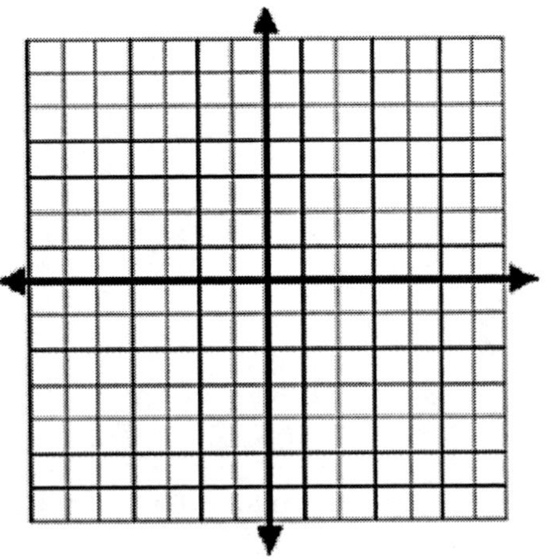

2. Graph $y = 3x + 1$

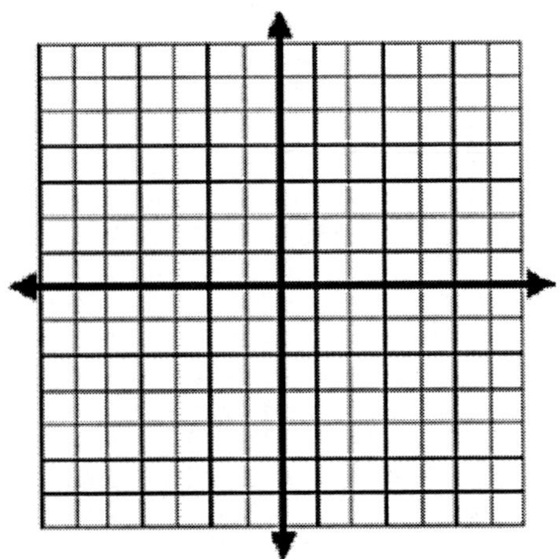

3. Graph $y = \frac{2}{3}x - 2$

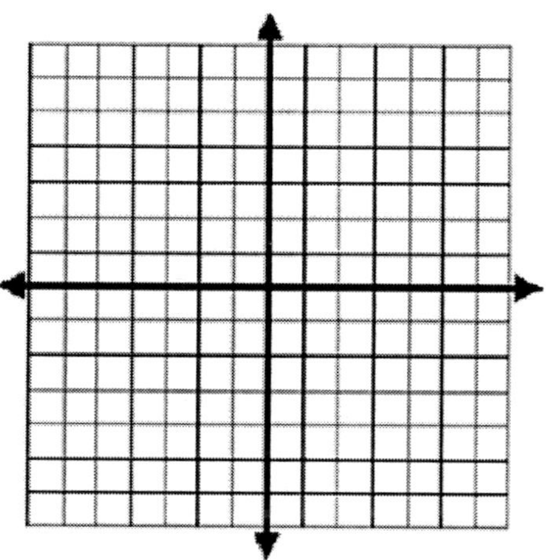

4. Graph on same graph: $y = -2, x = 3$

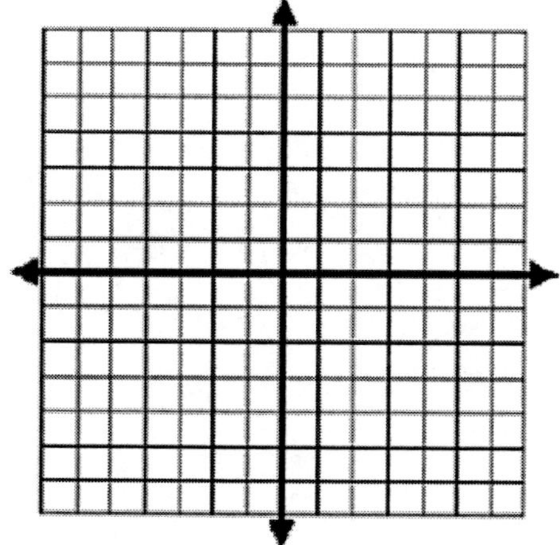

HOMEWORK ASSIGNMENT: 8.2

Name: ______________________________

1. Graph $y = -3x + 2$

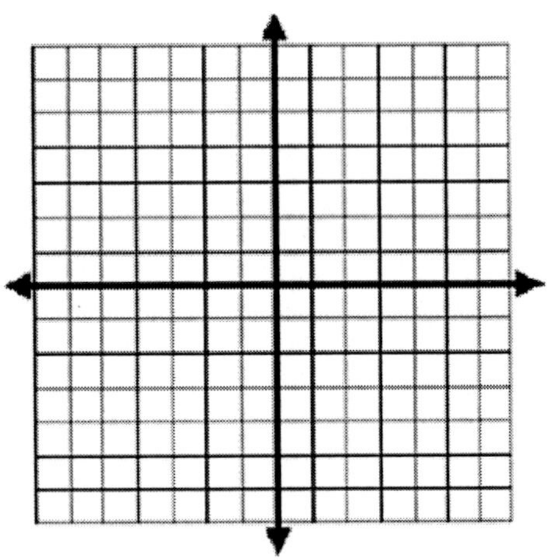

2. Graph $y = -\frac{1}{3}x + 2$

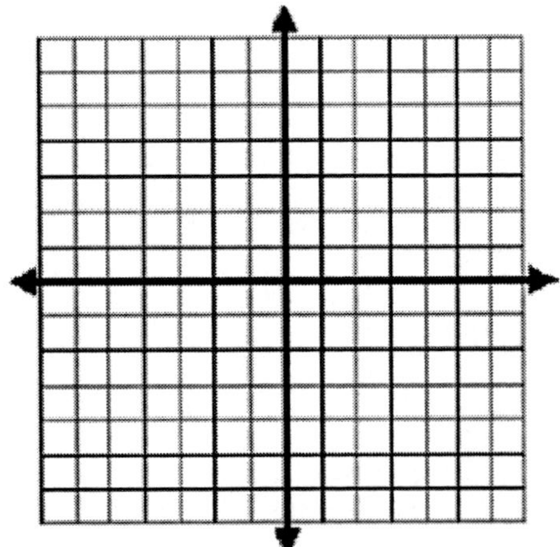

3. Graph $y = -2x - 5$

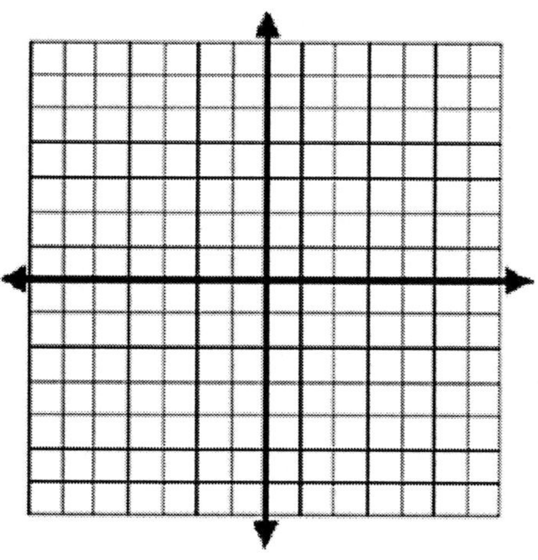

4. Graph $y = -\frac{2}{3}x - 2$

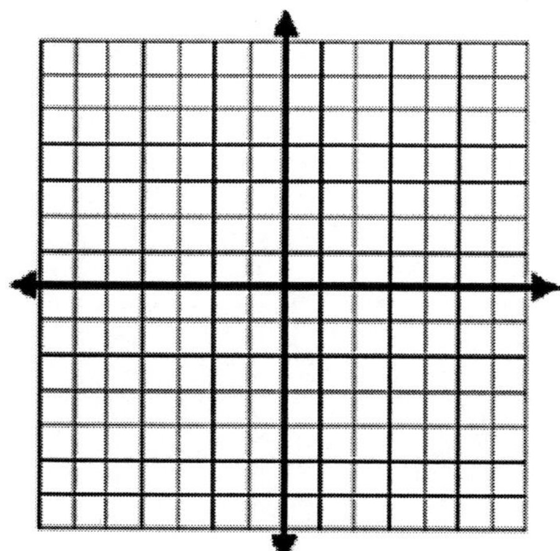

5. Graph $y = -\frac{5}{4}x + 3$

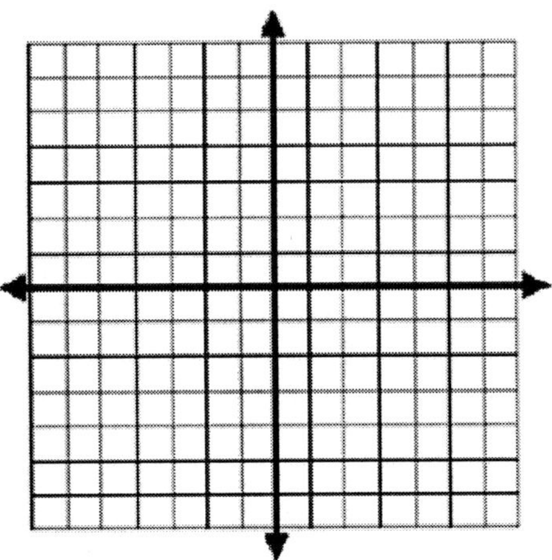

6. Graph $y = 3$, $x = -2$

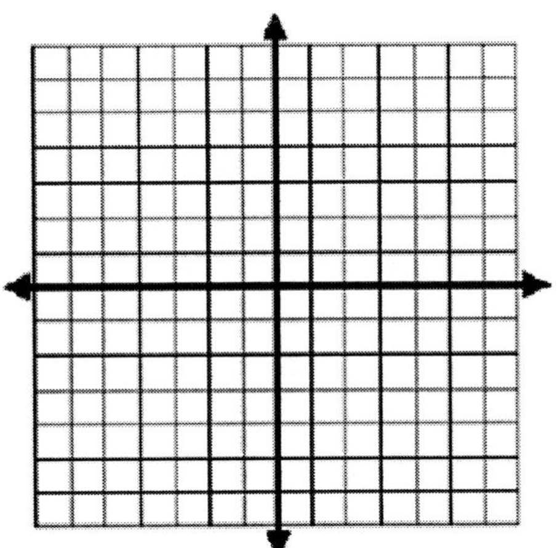

7. Graph $y = -4$, $x = -4$

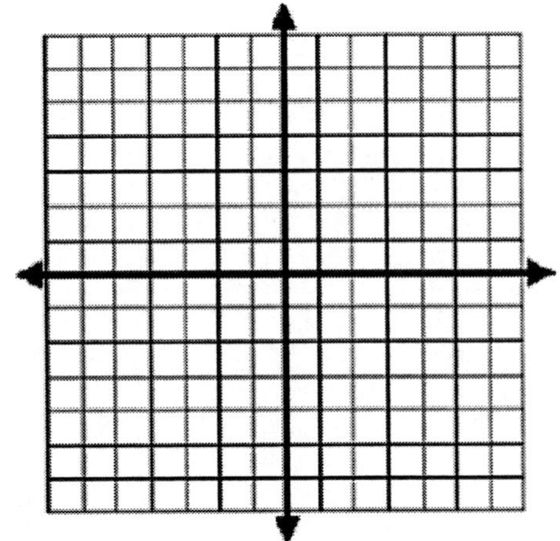

8.3 MULTIPLICATION RULES FOR EXPONENTS

Product Rule for Exponents

If the bases are the same and only differ in the exponents, keep the base and add the two exponents.

EXAMPLES:

1. $5^3 \cdot 5^6$
2. $m^2 \cdot m^3$
3. $b^3b^8b^2$
4. $4m \cdot 6m^5$
5. $-7k^5(-5k^3)(-2k^9)$
6. $-2a^4b^5(8a^2b^5)$
7. $-7y^5(5y^3)$
8. $-7k^5(-5k^3)(-2k^9)$
9. $-2a^4b^5(8a^2b^5)$

Power Raised to a Power: $(xm)^n = x^{m \cdot n}$

You must keep the base and just multiply the exponents; make sure you see parentheses to do the power of a power rule.

EXAMPLES:

1. $(4^2)^6$
2. $(y^6)^4$
3. $(x^4)^2 (x^3)^3$
4. $(5rs)^3$
5. $(6h^2s^9)2$
6. $(4y^3)^2 (3y^4)^3$
7. $(5x^3)^2 (5x^4)^3$
8. $(6t^5)^2 (2t^2)^2$
9. $(3ab^3)^2 (2a^4b^5)^3$

HOMEWORK ASSIGNMENT: 8.3

Name: ______________________________

1. $-6x^3(4x^2)$

2. $6y(2y^3)3y^4$

3. $s^2t \bullet st$

4. $w^2y \bullet yw^4$

5. $-2rt^4\,(-5r^2\,t^2)$

6. $(2s^2\,t^3)^2$

7. $(4h^5\,y^6)^2$

8. $(3a^3)^3\,(2a^2)^3$

9. $(3y^2y^5)^3$

8.4 INTRODUCTION TO POLYNOMIALS

Polynomial

A term or sum of terms in which all variables have whole number exponents.

Classifying polynomials

Monomial	Binomial	Trinomial
One term	Two terms	Three terms
$5x^2$	$2x - 1$	$5t^2 + 4t + 3$
$-6x$	$18a^2 - 4a$	$27x^3 - 6x + 2$
29	$-27z^4 + 7z^2$	32r2 + $7r - 12$

Degree of a term: of a polynomial in one variable is the value of the exponent on the variable. If a polynomial is in more than one variable, the degree of a term is the sum of the exponents on the variables. The degree of a nonzero constant is 0

Degree of a polynomial: the highest degree of any term of the polynomial.

EXAMPLE:

List the terms, the coefficients and the degree of each term, and the degree of the polynomial(Hint: Rearrange the polynomial in decreasing power)

1. $6p^3$

2. $17r^4 + 2r^8 - r$

3. $-2g^5\ 7g^6 + 12g^7$

4. $\frac{1}{2}p^4 - p^2$

Evaluate: x = −1

1. $-2x^2 - 4$

2. $3x^2 - 4 + 1$

Graphing Equations Involving Polynomials

Previously we graphed linear equations in the form: $y = mx + b$

The easiest way to graph second–degree polynomials in the form: $y = ax^2 + bx + c$ is to make a table of values. The best values to choose are −1, 0, 1. If a fraction is involved then choose the denominator's value in ± and 0.

FOR EXAMPLE:

$y = 3x^2$

X	Y
−1	
0	
1	

$y = 3x^2$

X	Y
−4	
0	
4	

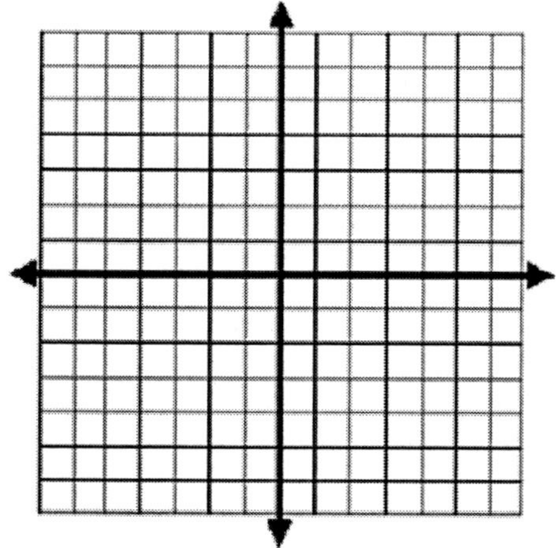

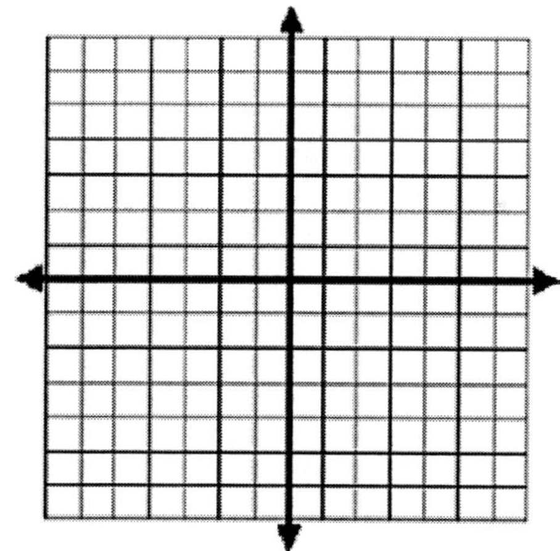

Applications:

1. The height h (in feet) of a ball shot straight up with an initial velocity of 64ft/s is given by the equation: $h = -16t^2 + 64t$. Find the height of the ball after the given number of seconds.

   a. 0 sec

   b. 1 sec

   c. 2 sec

   d. 4 sec

2. The number of feet that a car travels before stopping depends on the driver's reaction time and the braking distance. For one driver, the stopping distance d is given by the equation $d = 0.04v^2 + 0.9v$, where v is the velocity of the car. Find the stopping distance for each of the following speeds.

   a. 30mph

   b. 60mph

Review:

3. $\frac{5}{14} - \frac{4}{21}$

4. $\frac{23}{25} \div \frac{46}{5}$

5. $3(a - 5) = 4\ (a + 5)$

HOMEWORK ASSIGNMENT: 8.4

Name: ______________________________

Classify, list the terms, the coefficients and the degree of each term, and the degree of the polynomial(Hint: Rearrange the polynomial in decreasing power)

1. $5t^2 - t + 1$

2. $-3q + 2 - q^2$

Evaluate:

3. $\frac{2}{3}b^2 - b + 1$ *when* $b = 3$

4. $3n^2 - n + 2$ *when* $n = 2$

5. $-2s^2 - 2s + 1$ *when* $s = -1$

6. $-4r^2 - 3r - 1$ *when* $r = -2$

Graph:

7. $y = 2x^2 - 3$

X	Y
−1	
0	
1	

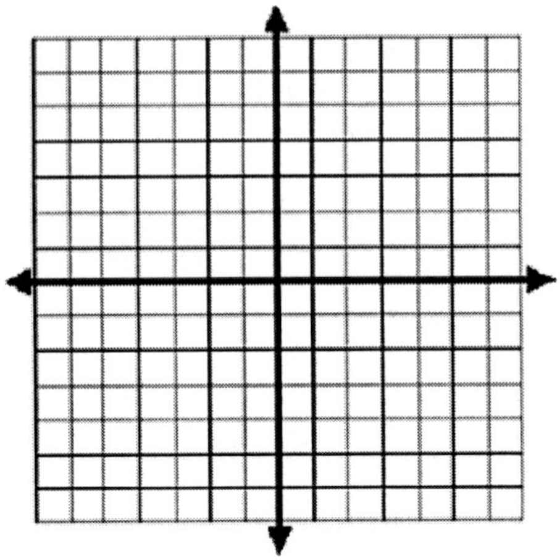

8. The number of feet that a car travels before stopping depends on the driver's reaction time and the braking distance. For one driver, the stopping distance d is given by the equation $d = 0.04v^2 + 0.9v^2$, where v is the velocity of the car. Find the stopping distance for each of the following speeds.

a. 50mph

b. 70mph

8.5 ADDING AND SUBTRACTING POLYNOMIALS

Adding polynomials is basically collecting like terms. We usually arrange the terms of a polynomial in descending order for a particular variable.

EXAMPLE:

1. $3.2m^3 + 4.5m^3 + 7.2m^3$

2. $(5y - 2) + (-3y + 7)$

2. $(2b^2 - 4b) + (b^2 + 3b - 1)$

4. $(s^2 + 1.2s - 5) + (3s^2 - 2.5s + 4)$

To subtract polynomials change all the signs inside the parentheses after the negative sign. (Hint: Distribute the negative one in–front of the parentheses before combing like terms.)

1. $6y^3 - 9y^3$

2. $(3.3a - 5) - (7.8a + 2)$

2. $(5y^2 - 4y + 2) + (3b^2 + 2y - 1)$

4. $(3x^2 + 4x - 5) + (-6y^2 - 6y - 13)$

EXTRA PRACTICE:

1. $(3x^2 + 2x - 4) + (5x^2 - 17)$

2. $(4c^2 + 3c - 2) + (3c^2 + 4c + 2)$

3. $(2b^2 + 3b - 5) - (2b^2 - 4b - 9)$

4. $-2x^2 - 4x + 12 - (10x^2 + 9x - 24)$

Applications:

5. If a house is purchased for \$85,000 and is expected to appreciate \$700/yr, its value y after x years is given by the equation: $y = 700x + 85{,}000$. Find the value after 10 years.

6. A young couple bought two cars, one for \$8,500 and the other for \$10,200. The first car is expected to depreciate \$800/yr and the second car \$1,100/yr.

   a. Write an equation that will give the value y of the first car after x years.

   b. Write an equation that will give the value y of the second car after x years.

   c. Find a single equation that will give the value y of both cars after x years.

   d. Find the value of both cars after 6 years.

HOMEWORK ASSIGNMENT: 8.5

Name: ______________________________

1. $(4p^2 - 4p + 5) + (6p - 2)$

2. $(3n^2 - 5.8n + 7) + (-n^2 + 5.8n - 2)$

3. $(-3t^2 - t + 3.4) + (3t^2 + 2t - 1.8)$

4. $$\begin{array}{l} -6x^3 - 4.2x^2 + 7 \\ \underline{+(-7x^3 + 9.7x^2 - 21)} \end{array}$$

5. $(3x^2 - 2x - 1) - (-4x^2 + 4)$

6. $(3a^2 - 2a + 4) - (a^2 - 3a + 7)$

7. $(m^2 - m - 5) - (m^2 + 5.5m - 7.5)$

8. $$\begin{array}{l} (3x^2 + 4x - 5) \\ \underline{-(-2x^2 - 2x + 3)} \end{array}$$

9. If a house is purchased for $85,000 and is expected to appreciate $700/yr, its value y after x years is given by the equation:

   $y = 700x + 85{,}000$. Find the value of the house after 16 years.

10. A ship entering the Panama Canal from the Atlantic Ocean is lifted up 85 feet to Lake Gatun by the Gatun Lock system. Then the ship is lowered 31 feet by the Pedro Miguel Lock. By how much must the ship be lowered by the Miraflores Lock system for it to reach the Pacific Ocean water level?

8.6 MULTIPLYING POLYNOMIALS

Polynomials Times a Monomial

Use the distributive property that has been introduced in the earlier chapters.

EXAMPLES:

1. $3y(5y^2 - 4y)$

2. $5x(3x^2 - 2x + 3)$

Multiplying Binomials and Trinomials

The easiest way is to use a Punnett Square.

This is where you can place the multiplication symbol (•)	This is the 1st term of the 1st binomial (A)	This is the second term of the 1st binomial (a)
This is the 1st term of the 2nd binomial (B)	Product of (A)(B)	Product of (B)(a)
This is the 2nd term of the 2nd binomial (b)	Product of (b)(A)	Product of (b)(a)

EXAMPLE:

1. $(3x - 2)(2x + 3)$

2. $(5x - 4)^2$

3. $(5x + 4)^2$

4. $(5r + 6)(2r - 1)$

5. $(x - 2)(3x^2 + 4x + 1)$

6. $(4y - 3)(3y^2 - 5y + 4)$

Applications:

1. Express the area of the rectangle below.

$(x + 2)ft$

$(x - 2)ft$

2. The height of a triangular sail is $4x$ feet, and the base b is $(3x - 2)$ft. Find the area of the sail. (Hint: $A \frac{1}{2} = bh$)

HOMEWORK ASSIGNMENT: 8.6

Name: ____

1. $(2a + 4)(3a - 5)$

2. $(5t - 1)2$

3. $(t - 1)(2t + 4)$

4. $(3r + 7)(2r - 9)$

5. $(x + 4)(-2x^2 + 7x - 3)$

6. $(3y - 4)(y^2 + 8y - 3)$

7. Express the area of the rectangle below.

$(x - 4)ft$

$(x + 7)ft$

8. The height of a triangular sail is $3x$ feet, and the base b is $(4x + 2)$ feet. Find the area of the sail. (Hint: $A = \frac{1}{2}(bh)$

CHAPTER 9

RATIOS AND PROPORTIONS

9.1 RATIOS

Ratio

Is the quotient of two numbers of the quotient of two quantities that have the same units. There are three ways to express a ratio.

EXAMPLE:

Express each ratio in the lowest terms.

1. ratio of 25 to 10

2. ratio of .3 to 1.2

3. 32:16

4. 2 feet to 1 yard

5. 12 ounces to 1 pound

Rates

Rate: is a quotient of two quantities with different units.

EXAMPLE:

1. The fastest growing flowering plant on record grew 12 feet in 14 days. What was the rate of grown in this period?

2. A total of 78 inches of snow fell in a 24 hour period. What was the rate of snowfall?

Unit Rate

Unit rate: is a rate is which the denominator is 1.

EXAMPLE:

1. A driver makes the 354-mile trip from Orlando to Miami in 6 hours. What was his motorist's rate?

2. Joan earns $436 per 40-hour week managing a dress shop. Find her hourly rate.

3. A fast food restaurant sells a 12-ounce cola for 72¢ and a 16-ounce for 99¢. Which is the better buy?

4. To heat a house for 30 days, a furnace burned 69 therms of natural gas. Find the rate of gas consumption in the therms per day.

Applications:

EXAMPLE 1:

A 6ft tall tent standing next to a cardboard box casts a 9 ft shadow. If the cardboard box casts a shadow that is 6 ft long then how tall is it?

EXAMPLE 2:

A map has a scale of 3cm: 18km. If Riverside and Smithville are 54km apart then they are how much apart on the map?

EXAMPLE 3:

A model home is 12 cm wide. If it was built with a scale of 3 cm : 4 m then how wide is the real house?

HOMEWORK ASSIGNMENT: 9.1

Name: ____

Express each ratio in the lowest terms.

1. 1.5:2.4

2. 0.9:0.6

3. 12 minutes to 1 hour

4. 4 inches to 2 yards

Write as a unit rate:

5. 245 presents for 35 children

6. 60 revolutions in 5 minutes

7. 64 feet in 6 seconds

8. Four of us donated a total of $272.00.

9. An 11,880-gallon tank can be emptied in 27 minutes. Find the rate of flow in gallons per minute.

10. A car's odometer reads 34,746 at the beginning of a trip. Five hours later, it reads 35,071. How far has the car traveled? What is the average rate of speed?

11. One car went 1,235 miles on 51.3 gallons of gasoline, and another went 1,456 miles on 55.78 gallons. Which car got the better gas mileage?

12. Ricardo worked for 37 hours to help insulate a hockey arena. For his work, he received $536.50.Find his hourly rate of pay.

13. A telephone booth that is 8 ft tall casts a shadow that is 4 ft long. Find the height of a lawn ornament that casts a 2 ft shadow.

14. A map has a scale of 2 in : 6 mi. If Clayton and Centerville are 10 in apart on the map then how far apart are the real cities?

15. A statute that is 12 ft tall casts a shadow that is 15 ft long. Find the length of the shadow that a 8 ft cardboard box casts.

9.2 PROPORTIONS

Proportion: Is a statement that two ratios are equal.

Reduce both fractions to the smallest form and see if they are the same

EXAMPLE:

Determine whether each equation is a proportion.

1. $\frac{3}{7}=\frac{9}{21}$

2. $\frac{8}{3}=\frac{18}{39}$

Solving Proportions

Use cross-multiplication across the equal sign to break-up the fractions and form a one variable equation, then solve for the variable.

EXAMPLES:

1. $\frac{15}{x}=\frac{20}{32}$

2. $\frac{6.7}{x}=\frac{33.5}{38}$

3. $\frac{3m-1}{2}=\frac{12.5}{5}$

Writing Proportions to Solve Problems

EXAMPLE:

1. If 9 tickets to a concert cost $112.50, how much will 15 tickets cost?

2. How many cups of sugar will be needed to make several cakes that will require a total of 25 cups of flour?

3. An N-scale model railroad caboose is 4 in long. If N scale is 169 feet to 12 inches, how long is the real caboose?

4. If you can buy one can of pineapple chunks for $2 then how many can you buy with $26?

5. Mina was planning a trip to Jamaica for her honeymoon. Before going, she did some research and learned that the exchange rate is $1 US dollar for $89.38 Jamaican dollars. How much Jamaican dollars would she get if she exchanged $15 US dollars?

HOMEWORK ASSIGNMENT: 9.2

Name: ______________________________

Solve.

1. $\frac{4}{x}=\frac{2}{8}$

2. $\frac{x}{8}=\frac{9}{2}$

3. $\frac{x+1}{5}=\frac{3}{15}$

4. $\frac{x-1}{7}=\frac{2}{21}$

5. $\frac{4-x}{13}=\frac{11}{26}$

6. $\frac{2x-1}{18}=\frac{9}{54}$

7. $\dfrac{\frac{1}{2}}{\frac{1}{5}} = \dfrac{x}{2\frac{1}{4}}$

8. A manager of a school cafeteria orders 750 pudding cups. What will the order cost if she purchases them wholesale, 6 cups for $1.75?

9. Out of a sample of 500 men's shirts, 17 were rejected because of crooked collars. How many crooked collars would you expect to find in a run of 15,000 shirts?

10. Shawana reduced the size of a rectangle to the height of 2 inches. What is the new width if it was originally 24 inches wide and 12 inches tall?

11. Mary reduced the size of a painting to a width of 3.3 inces. What is the new height if it was originally 32.5 inches tall and 42.9 inches wide?

12. Molly bought two heads of cabbage for \$1.80. How many heads of cabbage can she buy if she had \$28.80?

13. Alex earns \$412 for a 40-hour week. If he missed 10 hours of work last week, how much did he get paid?

14. Find $\frac{1}{2}$ of 520.

15. Will purchased a shirt on sale for \$17.50. Find the original cost of the shirt if it was marked down 30%.

9.3 SQUARE ROOTS

When we take square roots of numbers, just treat them as if you are doing a factor tree. Break them down to their prime number combinations and only take the pairings out.

When taking the square root of an equation like: , you get two answers, , because positive 5's multiplied together gives an answer of 25 and negative 5's multiplied together also gives an answer of 25.

EXAMPLES:

1. Square root of 64.

Simplify:

2. $\sqrt{144}$

3. $-\sqrt{81}$

Evaluate:

4. $\sqrt{121}+\sqrt{1}$

5. $-\sqrt{9}-\sqrt{16}$

6. $8\sqrt{64}$

7. $-6\sqrt{25}+2\sqrt{36}$

8. $\sqrt{\frac{16}{49}}$

9. $\sqrt{0.04}$

Applications:

1. Find the length of the slanted side of each roof truss.

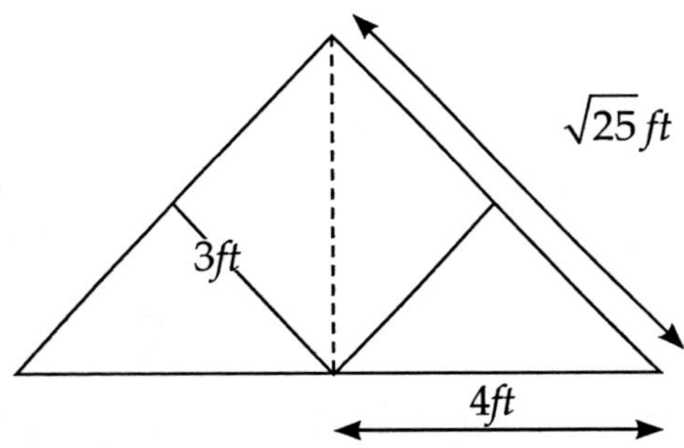

Review:

2. Simplify:

3. Evaluate: $5(-2)^2 - \frac{16}{4}$

4. Translate to mathematical symbols: four less than twice x.

5. Solve: $8 + \frac{a}{5} = 14$

HOMEWORK ASSIGNMENT: 9.3

Name: ____

Simplify.

1. $-\sqrt{\frac{1}{81}}$

2. $-\sqrt{\frac{64}{25}}$

3. $\sqrt{\frac{36}{121}}$

4. $2+6\sqrt{16}$

5. $-4\sqrt{49}+2\sqrt{25}$

6. $-6\sqrt{64}+5\sqrt{9}$

7. $\sqrt{\frac{1}{16}}-\sqrt{\frac{9}{25}}$

8. $5\left(\sqrt{81}\right)(-3)$

9. Find the length of the slanted side of each roof truss.

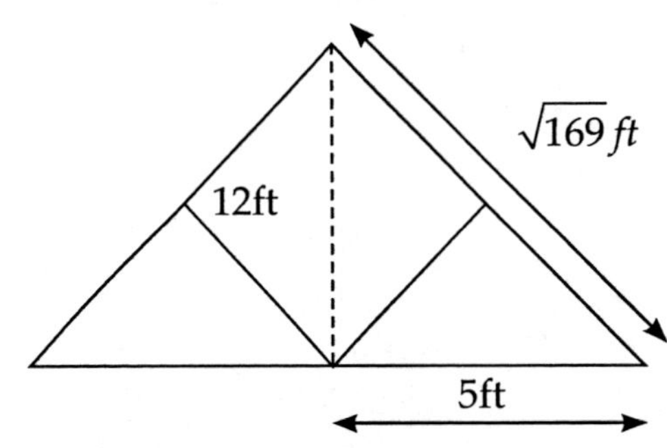

10. Place the following radicals in their respective locations on the number line below:
$\sqrt{1}, \sqrt{4}, \sqrt{16}, \sqrt{49}, \sqrt{81},$

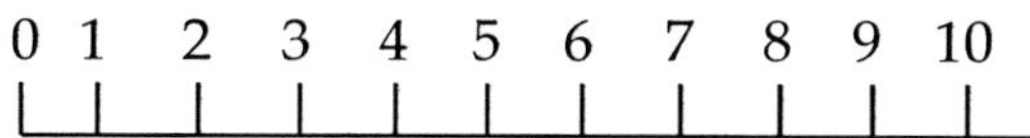

Review:

11 $5(-2)^2 - \frac{16}{4}$

12. $-(5x - 4) + 6(2x - 7) = -3$

13 The novel Fahrenheit 451, by Ray Bradbury, is a story about censorship and book burning. Use the formula: $C = \frac{5}{9}(F-32)$ to convert 451°F to degrees Celsius. Round your answer to the nearest tenth of a degree.

14. At a sports stadium, a food vendor is paid \$24 a game plus 30 for each bag of peanuts she sells. How many bags of peanuts must she sell to make \$60 a game?

9.4 AMERICAN UNITS OF MEASUREMENT

American Unit Length

1 foot = 12 in. 1 yard = 36 in. 1yard = 3 ft 1 mile = 5,280 ft

To convert from	Use the following	To convert from	Use the following
Feet to inches	$\frac{12\ in.}{1\ ft}$	Inches to feet	$\frac{1\ ft}{12\ in.}$
Yards to feet	$\frac{3\ ft}{1\ yd}$	Feet to yards	$\frac{1\ yd}{3\ ft}$
Yards to inches	$\frac{36\ in}{1\ yd}$	Inches to yards	$\frac{1\ yd}{36\ ft}$
Miles to feet	$\frac{5{,}280\ ft}{1\ mi.}$	Feet to miles	$\frac{1\ mi.}{5{,}280\ ft}$

EXAMPLES:

Convert.

1. 9 yards to feet

2. 1.5 feet to inches

3. $2\frac{2}{3}$ feet to inches

4. 2 miles to feet

5. 5 yards to feet

6. 44 inches to feet

American Units of Weights

1 pound = 16 ounces

1 ton = 2,000 pounds

To convert from	Use the following	To convert from	Use the following
Pounds to ounces	$\frac{16\ oz}{1\ lb}$	Ounces to pounds	$\frac{1\ lb}{16\ oz}$
Tons to pounds	$\frac{2{,}000\ lb}{1\ ton}$	Pounds to tons	$\frac{1\ ton}{2{,}000\ lb}$

EXAMPLES:

Convert.

1. 40 ounces to pounds

2. 60 pounds to ounces

3. 7,000 pounds to tons

4. 2.5 tons to ounces

5. 48,000 ounces to tons

6. 80 ounces to pounds

American Units of Capacity

1 cup = 8 fluid ounces

1 pint = 2 cups

1 quart = 2 pints

1 gallon = 4 quarts

To convert from	Use the following	To convert from	Use the following
Cups to ounces	$\frac{8\ fl\ oz}{1\ c}$	Ounces to cups	$\frac{1\ c}{8\ fl\ oz}$
Pints to cups	$\frac{2\ c}{1\ pt}$	Cups to pints	$\frac{1\ pt}{2\ c}$
Quarts to pints	$\frac{2\ pt}{1\ qt}$	Pints to quarts	$\frac{1\ qt}{2\ pt}$
Gallons to quarts	$\frac{4\ qt}{1\ gal}$	Quarts to gallons	$\frac{1\ gal}{4\ qt}$

EXAMPLE:

Convert.

1. How many pints are in 1 gallon?

2. 3 pints of milk to ounces

3. 3 gal to fluid ounces

4. 32 fluid ounces to pints

5. 3 quarts to pints

6. 2 quarts to fluid ounces

Units of Time

1 minute = 60 seconds

1 hour = 60 minutes

1 day = 24 hours.

To convert from	Use the following	To convert from	Use the following
Minutes to seconds	$\frac{60\ sec}{1\ min}$	Seconds to minutes	$\frac{1\ min}{60\ sec}$
Hours to minutes	$\frac{60\ min}{1\ hr}$	Minutes to hours	$\frac{1\ hr}{60\ min}$
Days to hours	$\frac{24\ hr}{1\ day}$	Hours to days	$\frac{1\ day}{24\ hr}$

EXAMPLE:

Convert.

1. A solar eclipse can last as long as 450 seconds. How many minutes is this?

2. 691,200 seconds to days

3. 7,200 minutes to days

4. A college student walks 11 miles in 155 minutes. To the nearest tenth, how many hours does he walk?

5. The astronauts of the Apollo 8 mission, which was launched on December 21, 1968, were in space for 147 hours. How many days did the mission take?

6. In 1935, Amelia Earhart became the first woman to fly across the Atlantic Ocean alone, establishing a new record for the crossing: 13 hours and 30 minutes. How many seconds is this?

HOMEWORK ASSIGNMENT: 9.4

Name: ______________________________

Convert the following:

1. $4\frac{2}{3}$ yards to feet

2. 15,840 feet to miles

3. 8 pounds to ounces

4. 2.5 tons to ounces

5. 20 quarts to gallons

6. 3 gallons to fluid ounces

7. 2,400 seconds to hours

8. 691,200 seconds to days

9. When Dan Marino of the Miami Dolphins retired, it was noted that his career passing total was nearly 35 miles. How many yards is this?

10. The Great Sphinx of Egypt is 240 feet long. How many inches is this?

11. Each student attending Osceola Elementary receives one pint of milk for lunch each day. If 575 students attend the school, how many gallons of milk are used each day?

12. A college student walks 11 miles in 155 minutes. To the nearest tenth, how many hours does he walk?

9.5 METRIC UNITS OF MEASUREMENT

Length

1 kilometer(km) = 1,000 meters	1 meter = $\frac{1}{1,000}$ kilometer
1 hectometer(hm) = 100 meters	1 meter = $\frac{1}{100}$ hectometer
1 dekameter(dam) = 10 meters	1 meter = $\frac{1}{10}$ dekameter
1 decimeter (dm) = $\frac{1}{10}$ meter	1 meter = 10 decimeters
1 centimeter (cm) = $\frac{1}{100}$ meter	1 meter = 100 centimeters
1 millimeter (mm) = $\frac{1}{1,000}$ meter	1 meter = 1000 millimeters

King Henry died while drinking chocolate milk.

EXAMPLES:

Convert.

1. 860 centimeters to meters

2. 5.3 meters to millimeters

3. 5.15 kilometers to centimeters

4. 65.78 km to dekameters

5. 0.0068 hm to kilometers

6. 0.125 m to millimeters

Mass

1 kilogram (kg) = 1,000 grams	1 gram = $\frac{1}{1,000}$ kilogram
1 hectogram (hg) = 100 grams	1 gram = $\frac{1}{100}$ hectogram
1 dekagram (dag) = 10 grams	1 gram = $\frac{1}{10}$ dekagram
1 decigram (dg) = $\frac{1}{10}$ gram	1 gram = 10 decigrams
1 centrigram (cg) = $\frac{1}{100}$ gram	1 gram = 100 centigrams
1 milligram (mg) = $\frac{1}{1,000}$ gram	1 gram = 1000 milligrams

EXAMPLES:

Convert.

1. 5 kilograms to grams

2. One brand name for Verapamil is Isoptin. If a bottle of Isoptin contains 90 tablets, each containing 200 mg of active ingredient, how many centigrams of active ingredient are in the bottle?

3. The net weight of a bottle of olives is 284 grams. Find the smallest number of bottles that must be purchased to have at least 1 kilogram of olives.

4. 500 mg to grams

5. 1000 kg to grams

6. 2 kg to centigrams

Capacity

1 kiloliter (kL) = 1,000 liters	1 liter = $\frac{1}{1,000}$ kiloliter
1 hectoliter (hL) = 100 liters	1 liter = $\frac{1}{100}$ hectoliter
1 dekaliter (daL) = 10 liters	1 liter = $\frac{1}{10}$ dekaliter
1 deciliter (dL) = $\frac{1}{10}$ liter	1 liter = 10 deciliters
1 centriliter (cL) = $\frac{1}{100}$ liter	1 liter = 100 centiliters
1 milliliter (mL) = $\frac{1}{1,000}$ liter	1 liter = 1000 milliliters
1 millimeter = 1cm^3 = 1 cc	1000 cc = 1000 mL = 1 liter

EXAMPLES:

Convert.

1. How many millimeters are in two 2-liter bottles of cola?

2. 500 cL to liters

3. 2000 cc to liters

4. 3 cc to millimeters

5. 400 liters to hectoliters

6. 3 deciliters to millimeters

HOMEWORK ASSIGNMENT: 9.5

Name: ____

Convert the following:

1. 73.2 m to dm

2. 0.125 m to mm

3. 65.78km to dam

4. 0.074cm to hm

5. 2kg to cg

6. 500mg to dag

7. 500mL to L

8. 400hL to cL

9. A can of Café Bustelo has a net weight of 133 grams. Find the smallest number of cans that must be packaged to have at least 1 metric ton of coffee. (Hint: 1 metric ton = 1,000kg)

10. Solve: $\frac{[illegible]}{5} - 3 = -3$

9.6 CONVERTING BETWEEN AMERICAN AND METRIC UNITS

Equivalent Lengths	
American to Metric	Metric to American
1 in. ≈ 2.54 cm	1 cm ≈ 0.3937 in.
1 ft ≈ 0.3048 m	1m ≈ 3.2808 ft
1 yd ≈ 0.9144 m	1m ≈ 1.0936 yd
1 mi ≈ 1.6093 km	1 km ≈ 0.6214 mi

Equivalent weights and masses	
American to Metric	Metric to American
1 oz ≈ 28.35 g	1 g ≈ 0.035 oz
1 lb ≈ 0.454 kg	1 kg ≈ 2.2 lb

Equivalent capacities	
American to Metric	Metric to American
1 floz ≈ 0.030 L	1 L ≈ 33.8 fl oz
1 pt ≈ 0.473 L	1 L ≈ 2.1 pt
1 qt ≈ 0.946 L	1 L ≈ 1.06 qt
1 gal ≈ 3.785 L	1 L ≈ 0.264 gal

Formulas for temperatures revisited:

$$C = \frac{5}{9}(F - 32)$$

$$F = \frac{9}{5}c + 32$$

1. Hot coffee is 110°F. To the nearest tenth of a degree, express this temperature in degrees Celsius.

2. The thermal springs in Hot SpringsNational Park in central Arkansas emit water as warm as 62°C. Change this temperature to degrees Fahrenheit.

CHAPTER

10

GEOMETRY

10.1 SOME BASIC DEFINITIONS

Points, Lines, and Planes

Point: a geometric figure that has a position but no length, width, or depth.

Line: infinitely long with no stopping points in either direction with arrows at each end.

Plane: a flat surface that has length and width but no depth (a wall, ceiling, floor, etc).

Ray: the part of a line segment that begins at some point and continues forever in one direction with only one arrow.

Line segment: the part of a line that consists of points A and B and everything in between, has stopping points known as endpoints.

Point

Line

Plane

Line segment

Ray

Angles

Angle: a figure formed by two rays with a common endpoint. The common endpoint is called the vertex, and the rays are called sides.

One unit of measurement of an angle is the degree. Here is a Protractor which is the instrument used to measure the angles in degrees.

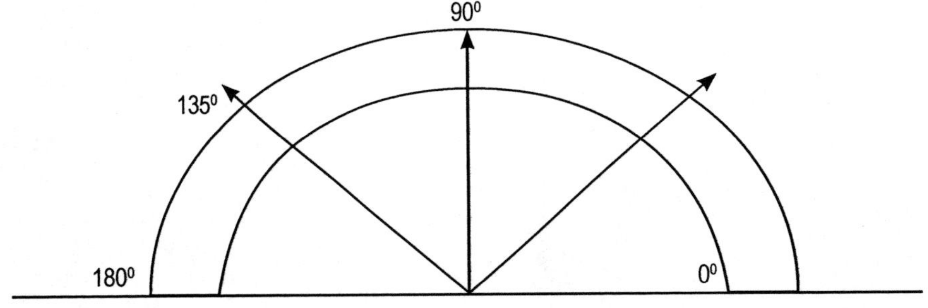

Classifications of Angles

Acute Angle: angles whose measures are greater than 0° but less than 90°	
Right Angle: angles whose measures are exactly 90°	
Obtuse Angle: angles whose measures are greater than 90° but less than 180°	
Straight Angle: angles whose measures are exactly 180°	

Adjacent Angles

two angles that are beside each other sharing a common ray line.

Vertical Angles

When two lines intersect, pairs of nonadjacent angles are called vertical angles. This really means that opposite angles as in the following example have the same measurements:

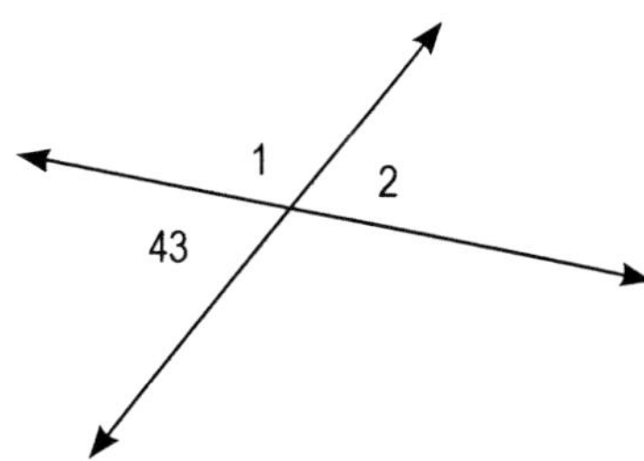

EXAMPLES:

Find the missing angles

1.

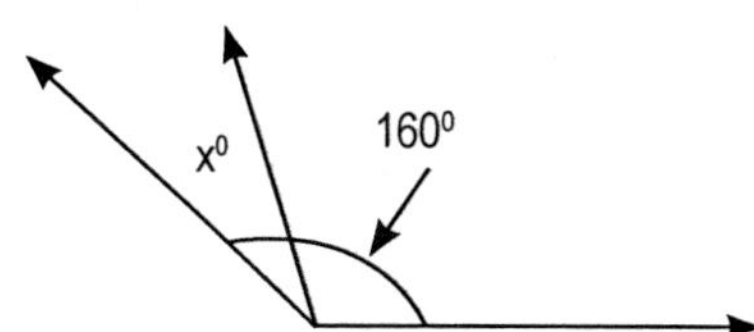

2.

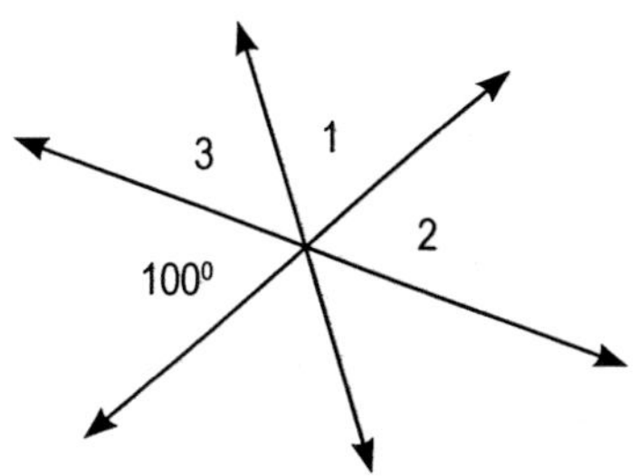

3.

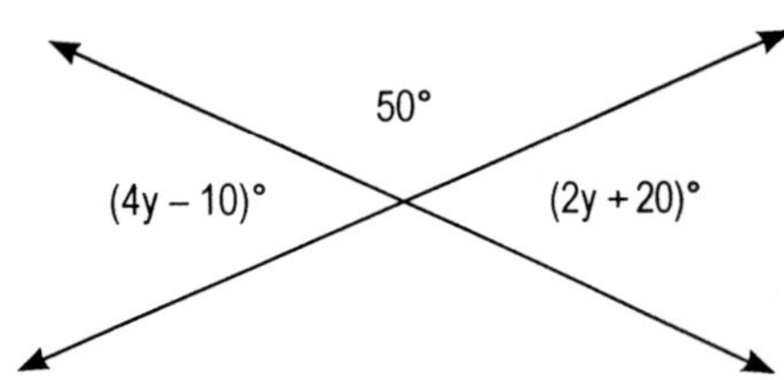

Complementary and Supplementary Angles

Two angles that have a common vertex and are side–by–side are called **adjacent angles**.

Two angles are **complementary angles** when the sum of their measures is 90°

Two angles are **supplementary angles** when the sum of their measures is 180°

EXAMPLES:

1. If the sum of the adjacent angles is 125°, what is the value of x.

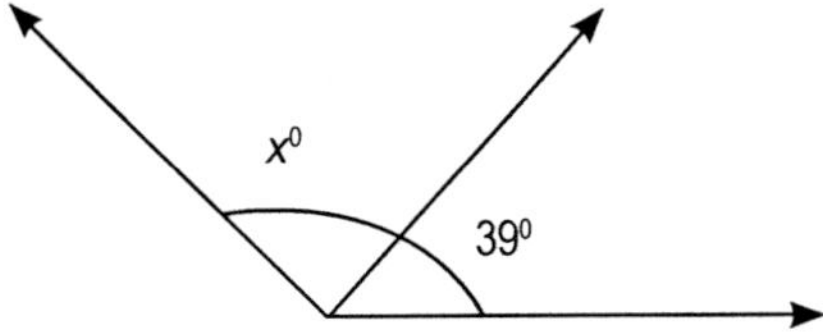

2. If the sum of the following supplementary angles is 180°, find the missing angle.

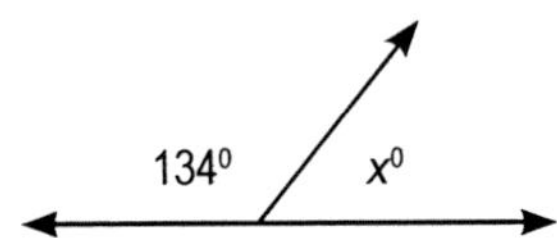

3. Find the complement of a 37° angle.

4. Find the supplement of a 117° angle.

HOMEWORK ASSIGNMENT: 10.1

Name: ______________________________

Find the value of x:

1.

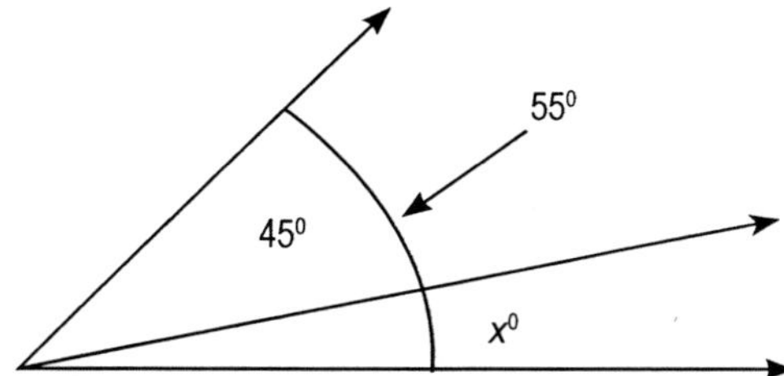

2.

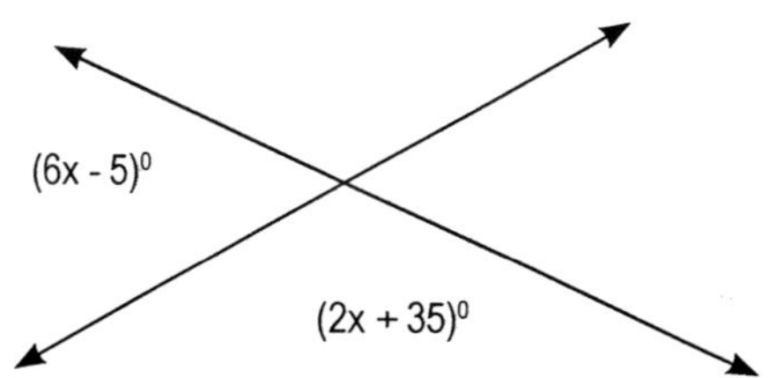

Let represent the unknown angle measure. Draw a diagram, write an appropriate equation, and then solve for .

4. Find the supplement of a 105° angle.

5. Find the complement of a 75° angle.

Review:

6. Subtract: $\frac{3}{4}-\frac{1}{8}-\frac{1}{3}$

7. Multiply: $\frac{5}{8}\cdot\frac{2}{15}\cdot\frac{6}{5}$

8. Graph: $y=-3x+5$

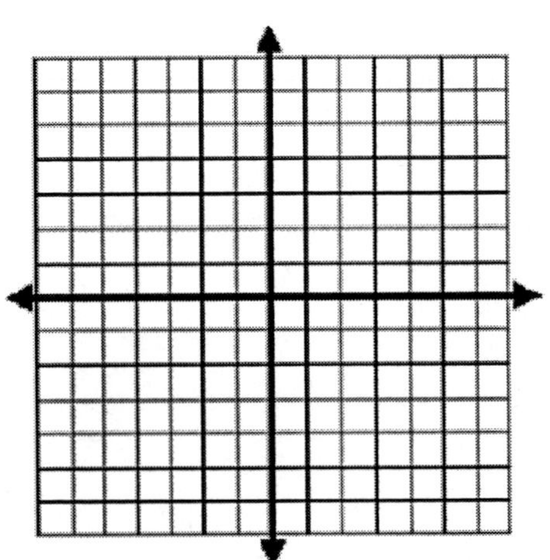

10.2 PARALLEL AND PERPENDICULAR LINES

Parallel and Perpendicular Lines

Parallel lines: are coplanar lines that do not intersect.	
Perpendicular lines: are lines that intersect and form right angles.	

Transversals and Angles

A line that intersects two or more coplanar lines is called a transversal.

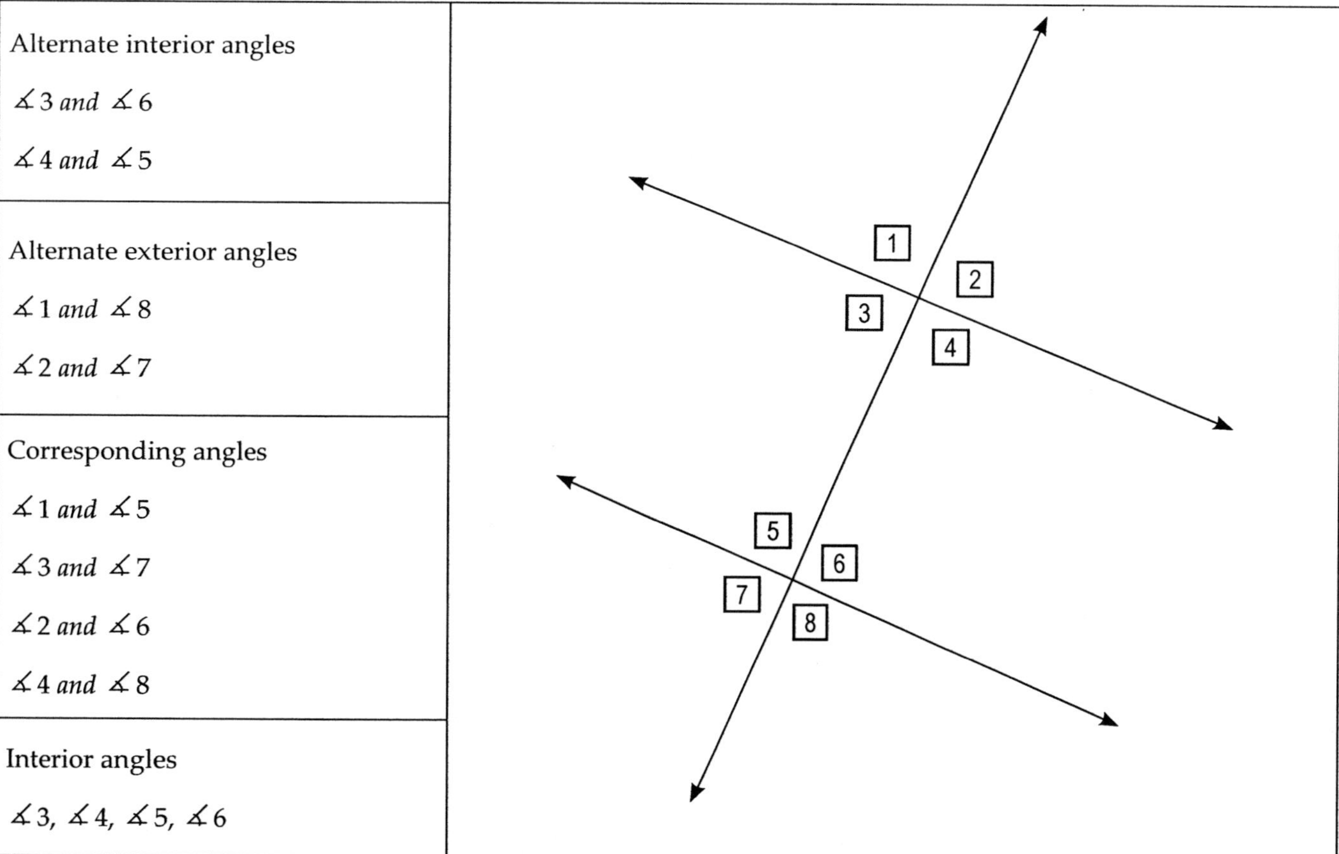

Alternate interior angles $\measuredangle 3$ *and* $\measuredangle 6$ $\measuredangle 4$ *and* $\measuredangle 5$	
Alternate exterior angles $\measuredangle 1$ *and* $\measuredangle 8$ $\measuredangle 2$ *and* $\measuredangle 7$	
Corresponding angles $\measuredangle 1$ *and* $\measuredangle 5$ $\measuredangle 3$ *and* $\measuredangle 7$ $\measuredangle 2$ *and* $\measuredangle 6$ $\measuredangle 4$ *and* $\measuredangle 8$	
Interior angles $\measuredangle 3$, $\measuredangle 4$, $\measuredangle 5$, $\measuredangle 6$	

Properties of Parallel Lines

EXAMPLE:

Find the missing angles if $\measuredangle = 120°$

Properties
Alternate interior angles are congruent $\measuredangle 3 \cong \measuredangle 6$ $\measuredangle 4 \cong \measuredangle 5$
Alternate exterior angles are congruent $\measuredangle 1 \cong \measuredangle 8$ $\measuredangle 2 \cong \measuredangle 7$
Corresponding angles are congruent $\measuredangle 1 \cong \measuredangle 5$ $\measuredangle 3 \cong \measuredangle 7$ $\measuredangle 2 \cong \measuredangle 6$ $\measuredangle 4 \cong \measuredangle 8$
Interior angles on the same side of the transversal are supplementary $\measuredangle 3 \cong \measuredangle 6 = 180^0$ $\measuredangle 4 \cong \measuredangle 6 = 180^0$

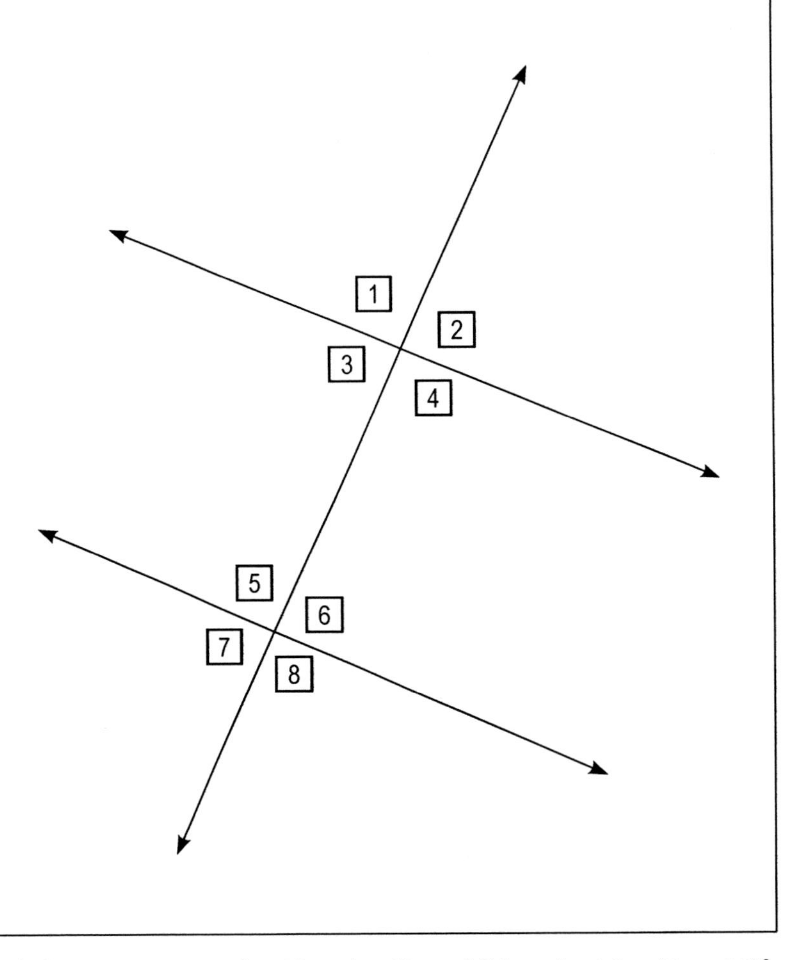

Using the figure in the previous example find the missing angles if $\measuredangle 4 = (2x + 10)^0$ and $\measuredangle 6 = (4x + 10)^0$. (Hint: Solve for the value of x and then find the measurement of either

HOMEWORK ASSIGNMENT: 10.2

Name: ____

1. In the figure line $\ell \parallel \overrightarrow{AB}$ Find the measure of each angle.

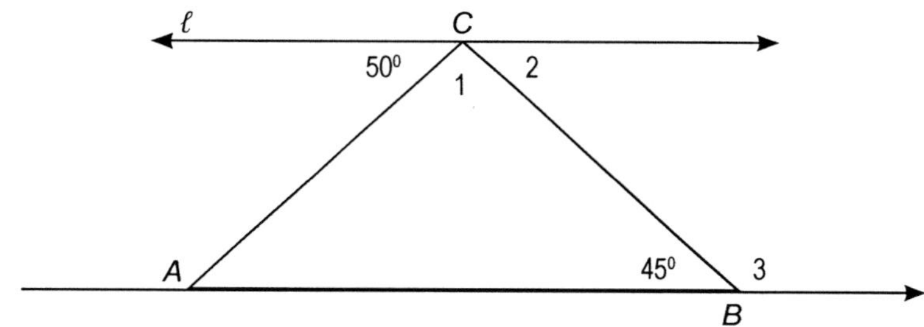

2. $\ell 2 \parallel \ell 2$ Find the value of .

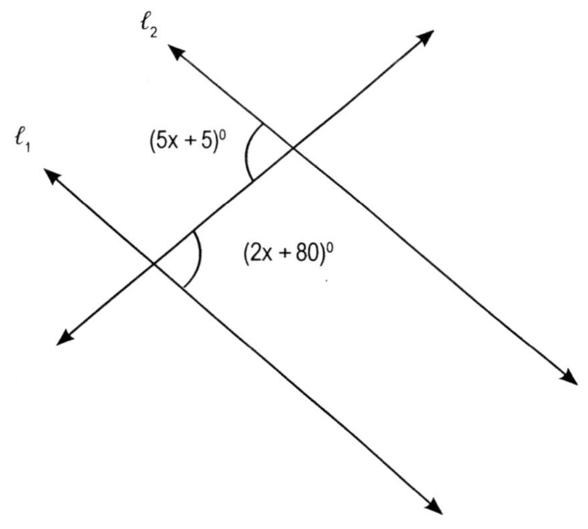

3.

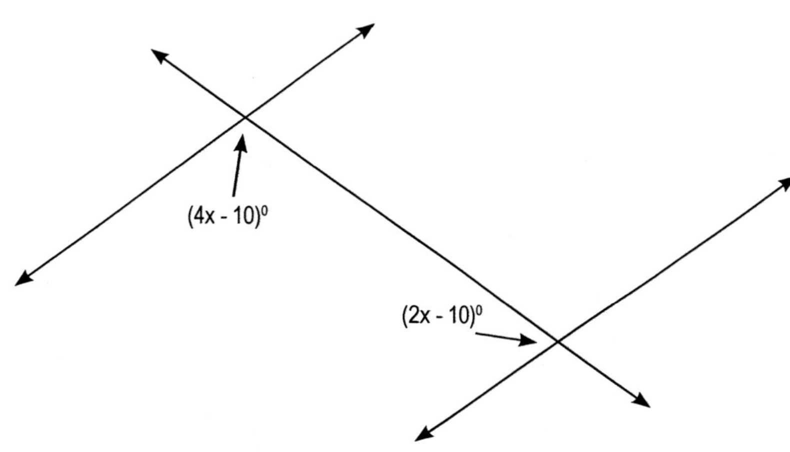

Review:

4. Find 60% of 120.

5. 80% of what number is 400?

6. Solve the proportion:

$$\frac{x+1}{18} = \frac{12.5}{45}$$

10.3 POLYGONS

A polygon is a closed geometric figure with at least three line segments for its sides.

	Triangle 3 sides	Quadrilateral 4 sides	Pentagon 5 sides	Hexagon 6 sides	Octagon 8 sides
Polygons					
Regular Polygons					

Triangles

Equilateral triangle: all sides of equal length.	
Isosceles triangle: at least two sides of equal length.	
Scalene triangle: no sides of equal length	
Right triangle: has one right angle.	

Sum of the Measures of a Triangle

The sum of the three angle measurements of *__any triangle__* is 180°.

EXAMPLES:

1. Find y.

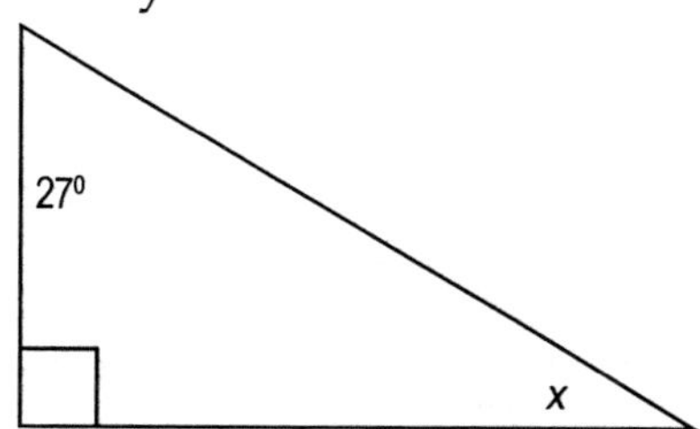

2. Find x.

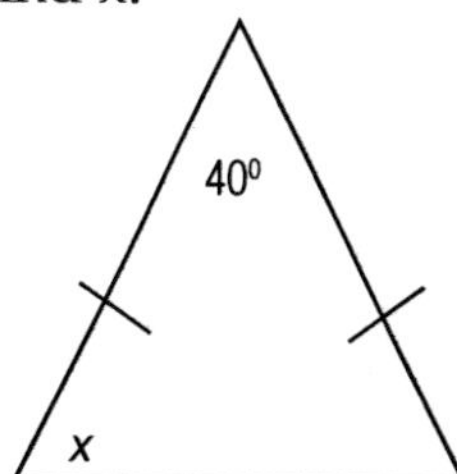

Quadrilaterals

A quadrilateral is a polygon with four sides. Some common quadrilaterals are shown below:

Parallelogram	Rectangle	Square	Rhombus	Trapezoid
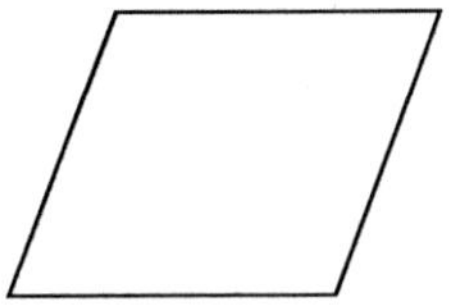	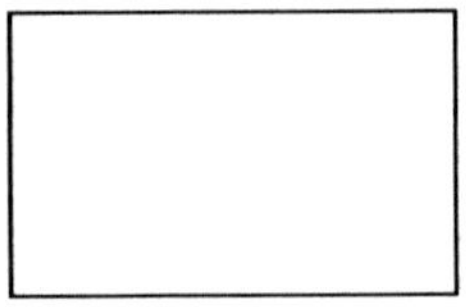		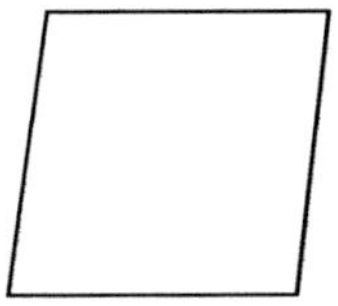	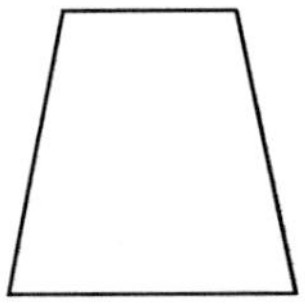

The Sum of the Measures of the Angles of a Polygon

The sum of the measures of the angles of a polygon with (n) sides is given by the formula: , or if you draw lines from one vertex of the figure repeatedly until you can not draw another line, then just multiply the number of triangles formed by 180°.

EXAMPLE:

1. Find the sum of the angles measures of a pentagon.
2. The sum of the measures of the angles of a polygon is 1,080°. Find the number of sides the polygon has.

HOMEWORK ASSIGNMENT: 10.3

Name: ______________________________

The measures of two angles of are given for the following figure. Find the measure of the 3rd angle.

1. $m\angle A = 30^0$, $m\angle B = 60^0$, $m\angle C =$ ______

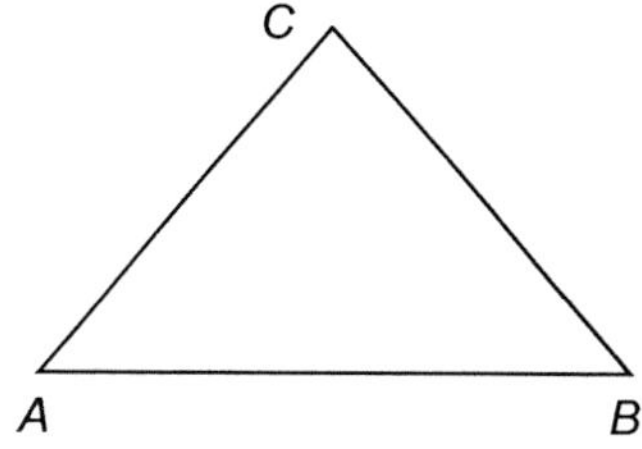

2. $m\angle B = 33^0$, $m\angle C = 77^0$, $m\angle A =$ ______

Refer to rectangle shown below:

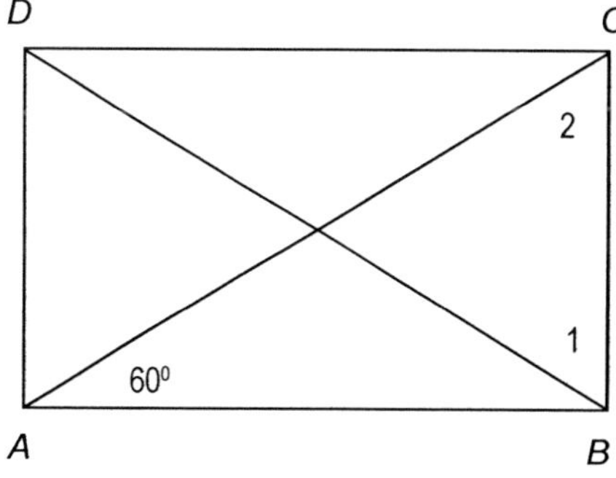

3. $m\angle 1 =$ ______

4. $m\angle 3 =$ ______

5. $m\angle 2 =$ ______

Review:

6. Find 20% of 110.

7. Find 15% of 50.

8. Simplify: $0.85 \div 2(0.25)$

9. When checking an accident victim's pulse, a paramedic counted 13 beats during a 15–second span. How many beats would be expected in 60 seconds?

10. $\frac{3}{2z-1}=\frac{3}{5}$

10.4 PROPERTIES OF TRIANGLES

Congruent Triangles

Two triangles whose sides and angles are the same.

Similar Triangles

All pairs of corresponding sides are in proportion and their angles are congruent.

EXAMPLE 1:

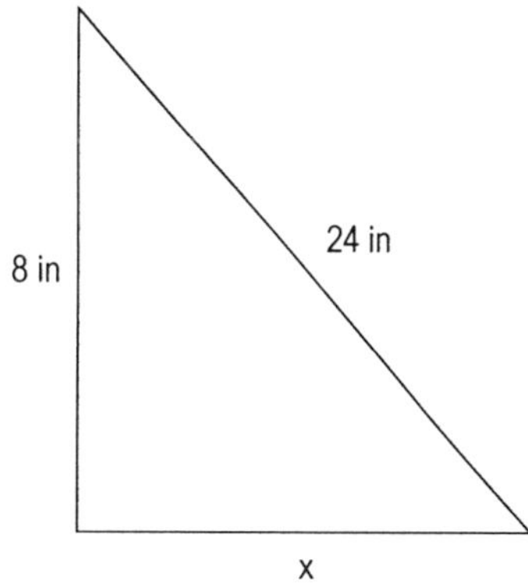

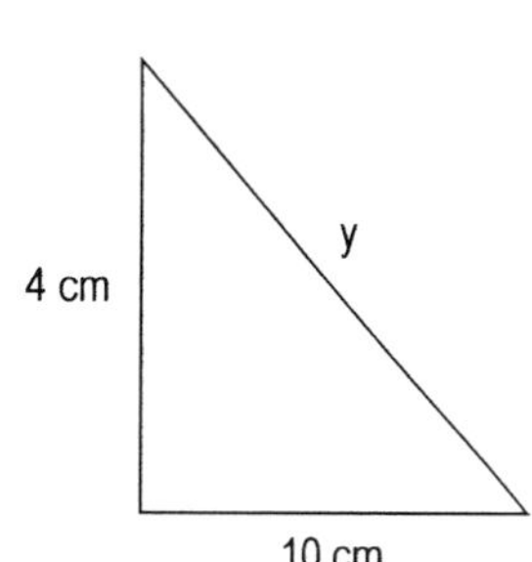

EXAMPLE 2:

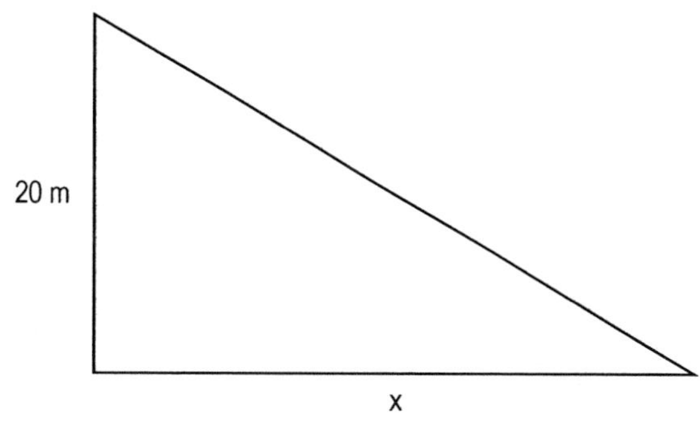

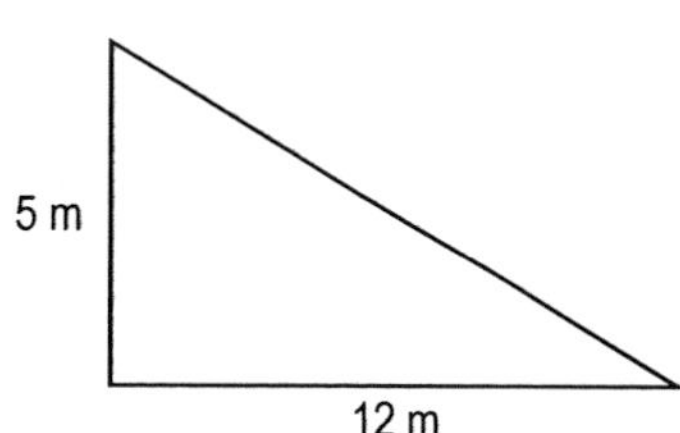

Pythagorean Theorem

In a right triangle, the sum of the squares of the lengths of the two shorter sides (legs) is equal to the square of the length of the longest side (hypotenuse).

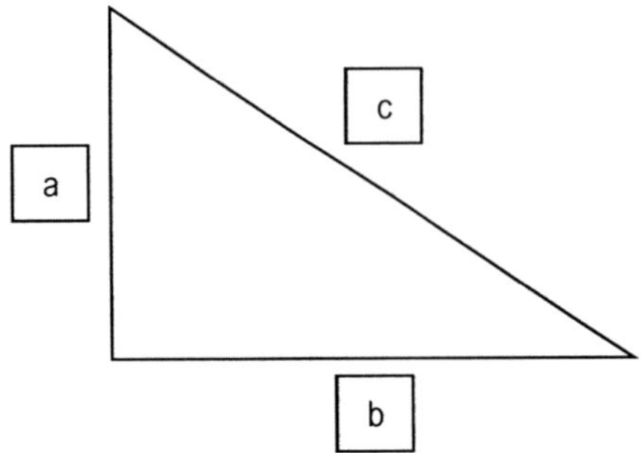

$a^2 + b^2 = c^2$

EXAMPLES:

Find the length of the missing side.

1.

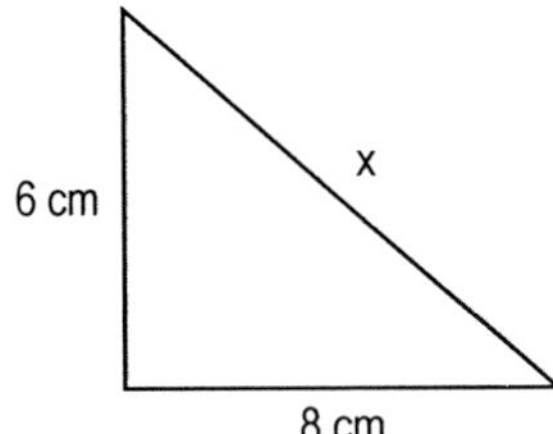

2.

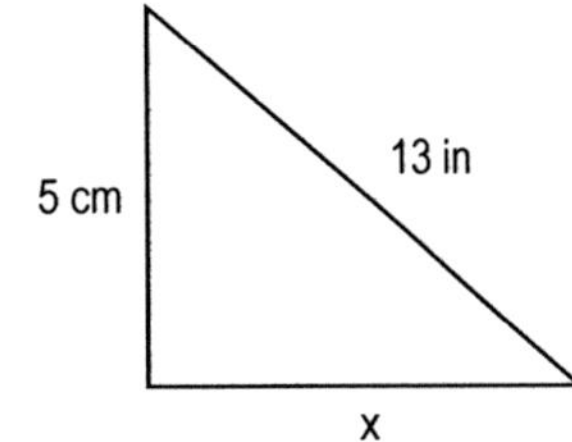

HOMEWORK ASSIGNMENT: 10.4

Name: ____________________

Solve for the missing letter:

1. a = 3 and b = 4. Find c.

2. a = 12 and b = 5. Find c.

3. a = 15 and c = 17. Find b.

4. b = 8 and c = 10. Find a.

5. A tree casts a shadow 24 feet long when a man 6 feet tall casts a shadow 4 feet long. Find the height of the tree. (Hint: Sketch a picture of the given information and use proportions)

6. Josie wants to find the height of the tallest building in Orlando. A line of sight from a point at ground level 456 feet away from the building lines up the top of a tree and the top of the building. The tree is 21 feet in front of her and it is 37 feet tall. How tall is the building to the nearest foot?

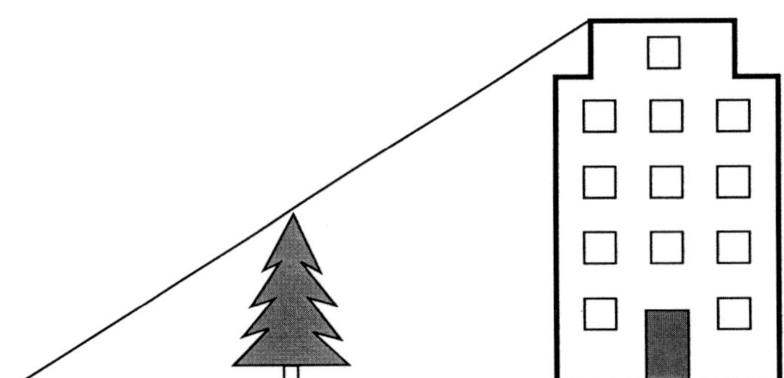

10.5 PERIMETERS AND AREAS OF POLYGONS

The perimeter is the distance around any shape.

The area is the measure of the amount of surface it encloses.

Area Formulas

Shape and formula	Figure labels
Square $A = s^2$	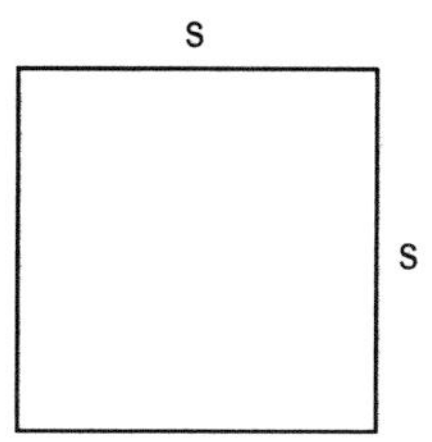 s, s
Rectangle $A = l \cdot w$	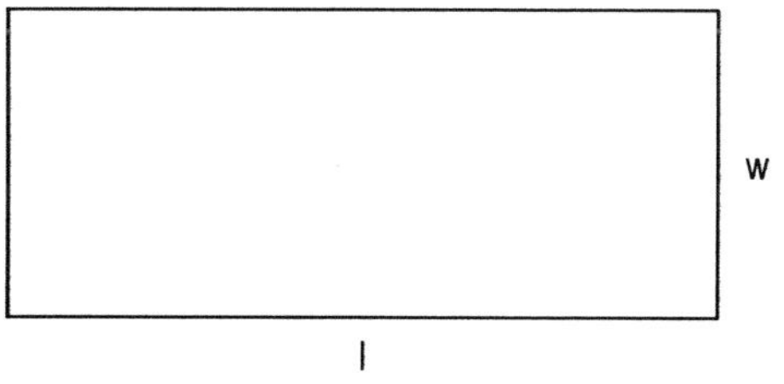 w, l
Parallelogram $A = b \cdot h$	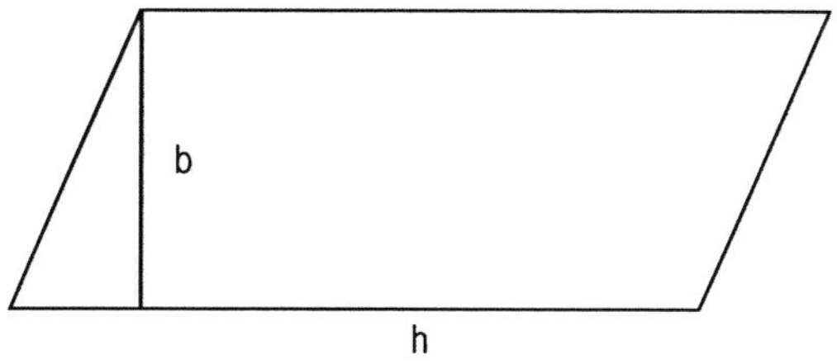 b, h
Triangle $A = \frac{b \cdot h}{2}$	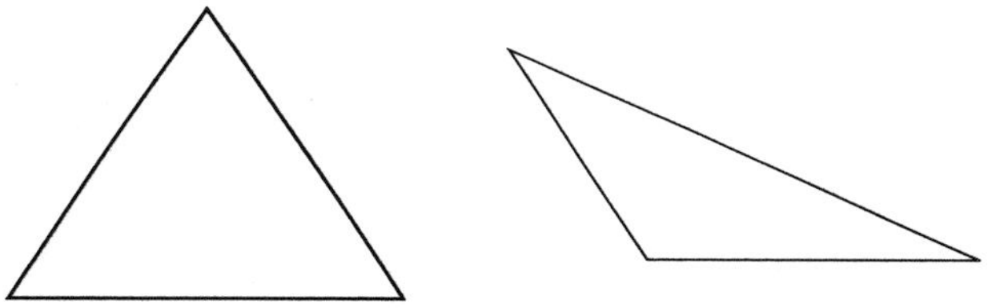
Trapezoid $A = \frac{h(b_1 + b_2)}{2}$	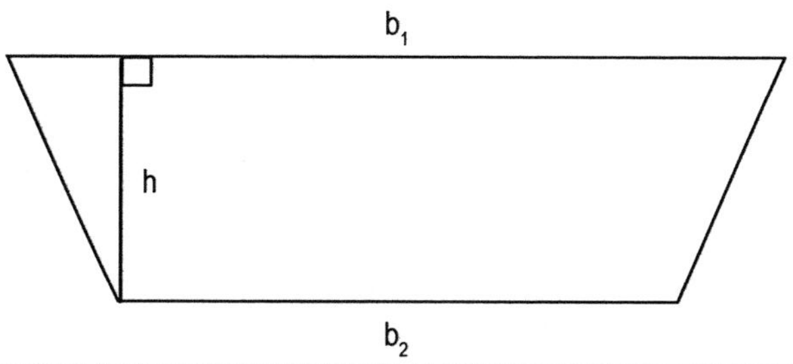 b_1, h, b_2

EXAMPLE:

Find the area and perimeter if possible of the following shapes.

1.

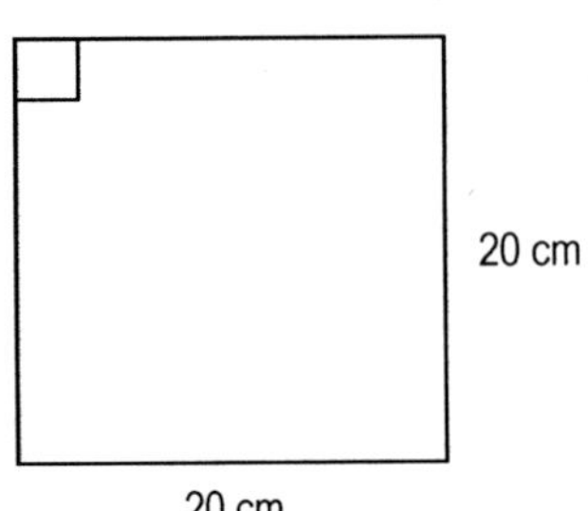

2.

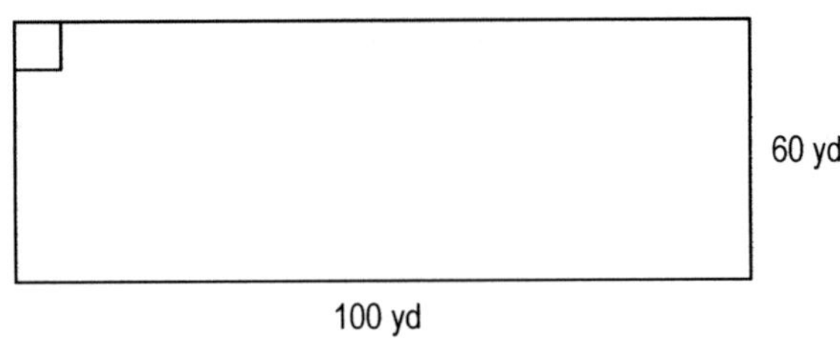

3.

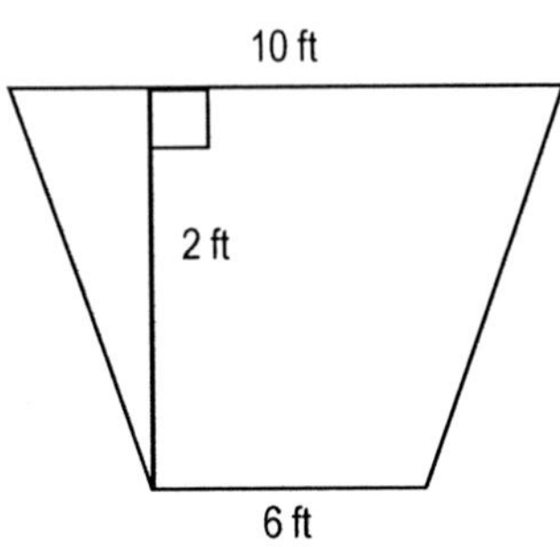

4.

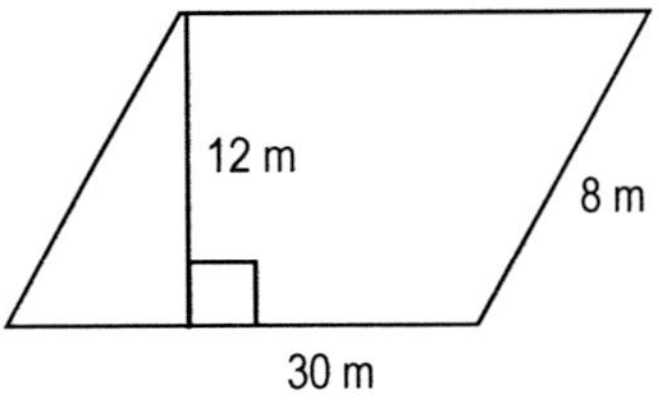

5.

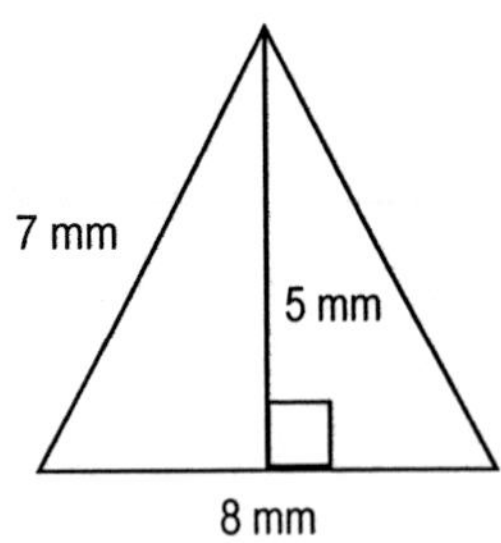

6.

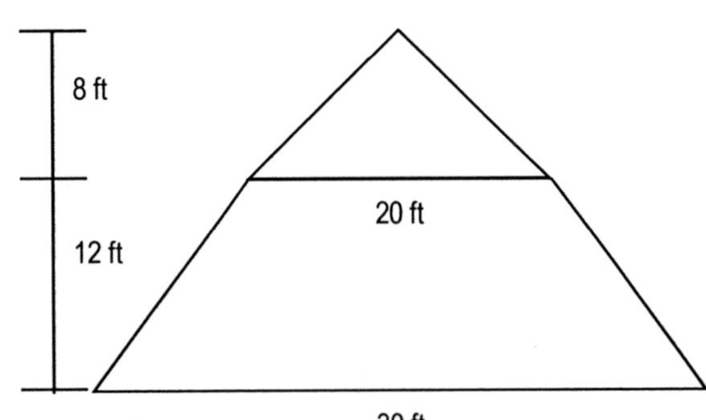

HOMEWORK ASSIGNMENT: 10.5

Name: ____

Find the area and perimeter of the following shapes:

1.

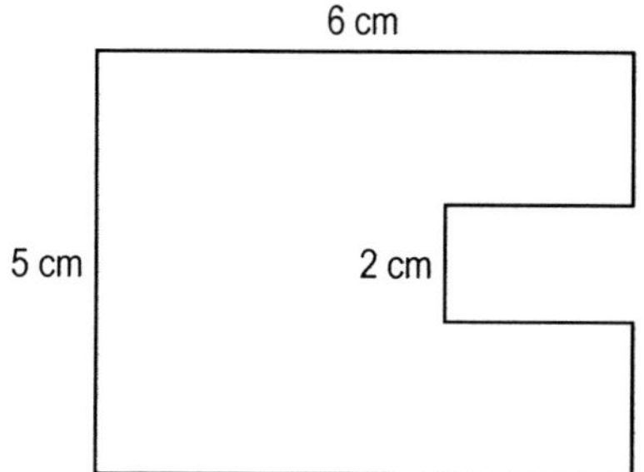

2.

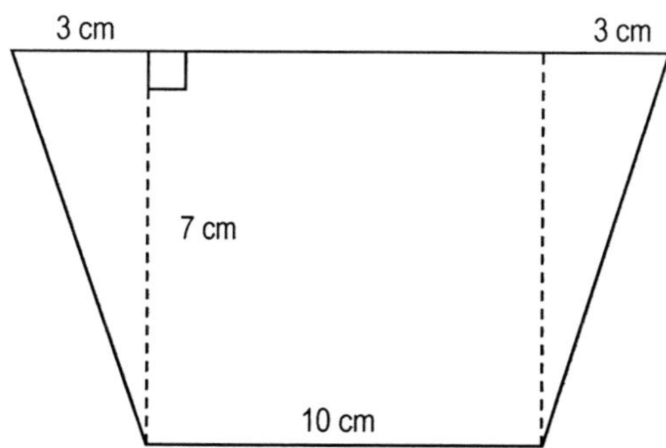

3.

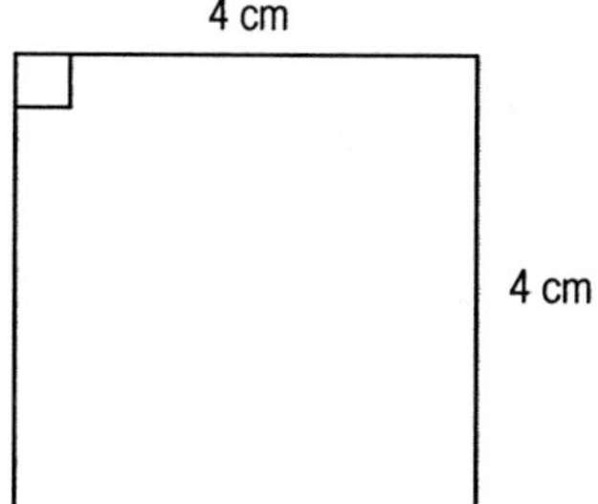

4.

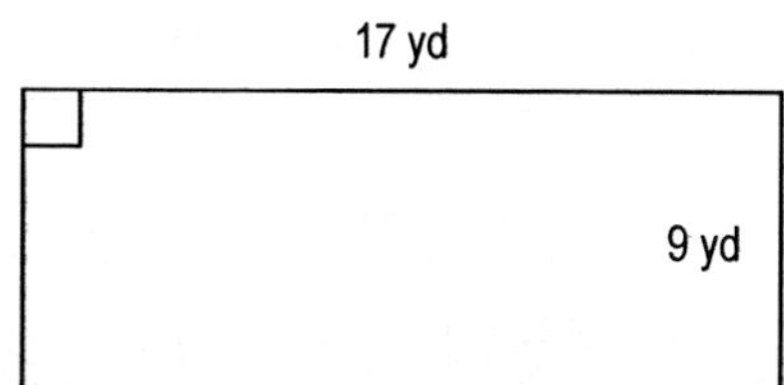

5. The length a rectangular garden is eight more than the width. The perimeter of the rectangle is 40 ft. What are the dimensions of the garden?

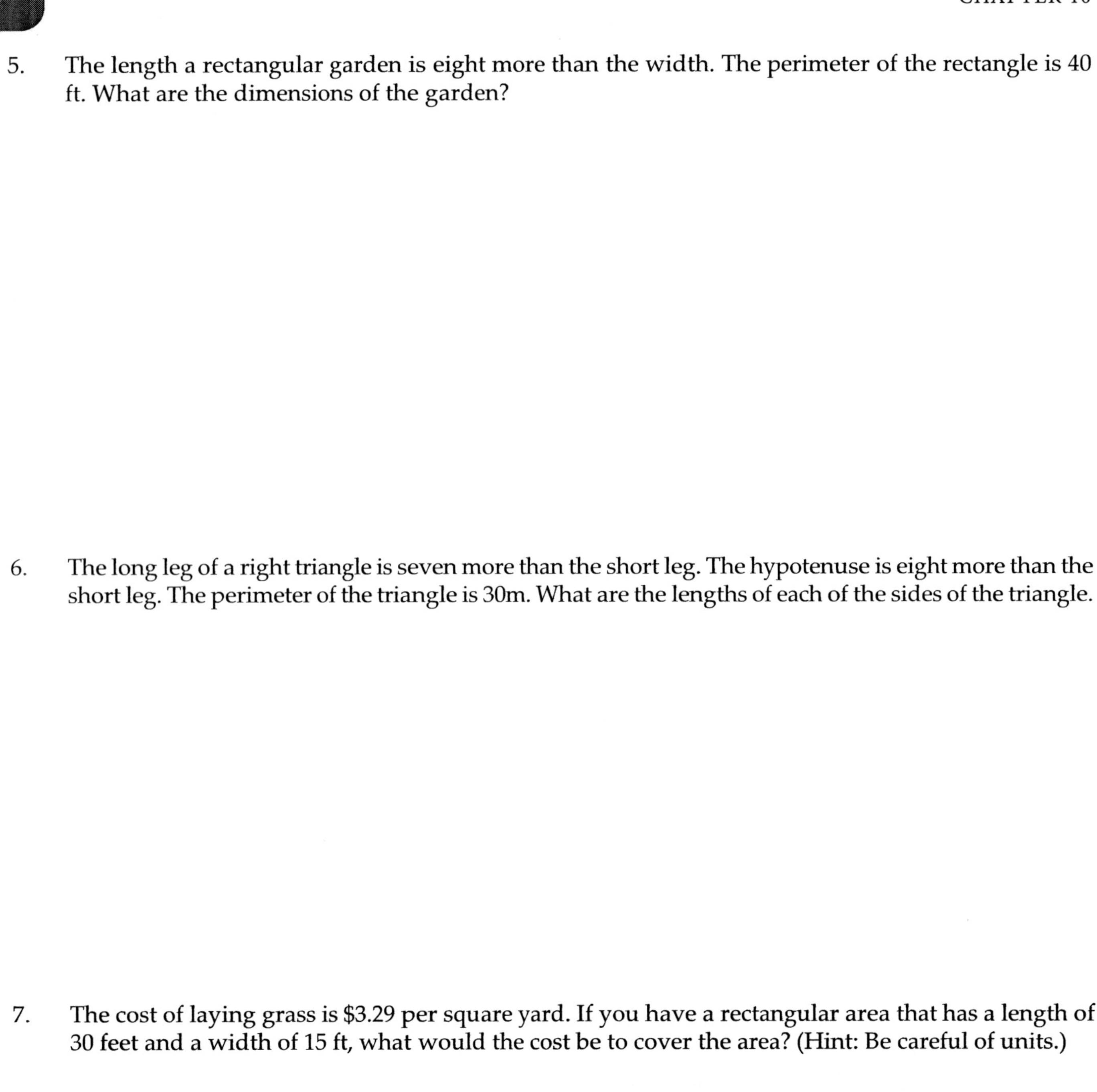

6. The long leg of a right triangle is seven more than the short leg. The hypotenuse is eight more than the short leg. The perimeter of the triangle is 30m. What are the lengths of each of the sides of the triangle.

7. The cost of laying grass is $3.29 per square yard. If you have a rectangular area that has a length of 30 feet and a width of 15 ft, what would the cost be to cover the area? (Hint: Be careful of units.)

10.6 CIRCLES

A circle is the set of all points in a plane that lie a fixed distance from a point called its center.

Diameter is the distance across the circle through the center. Radius is the distance from the center of the circle to any point on the circle.

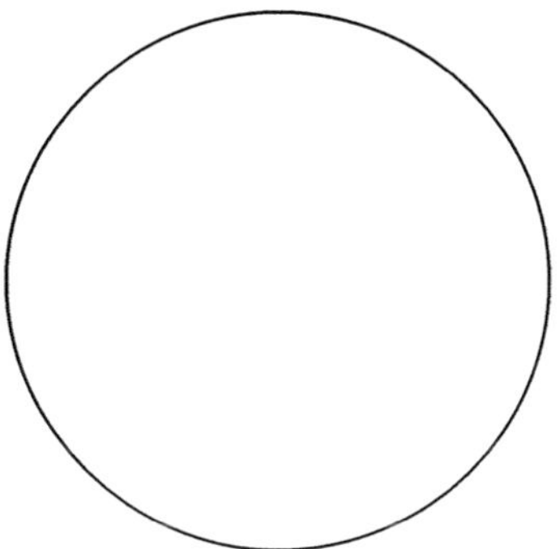

Circumference and Area of a Circle

$C = \pi d$ $A = \pi r^2$ $\pi \approx 3.14$

EXAMPLE:

Find the area and circumference of the following circles.

1.

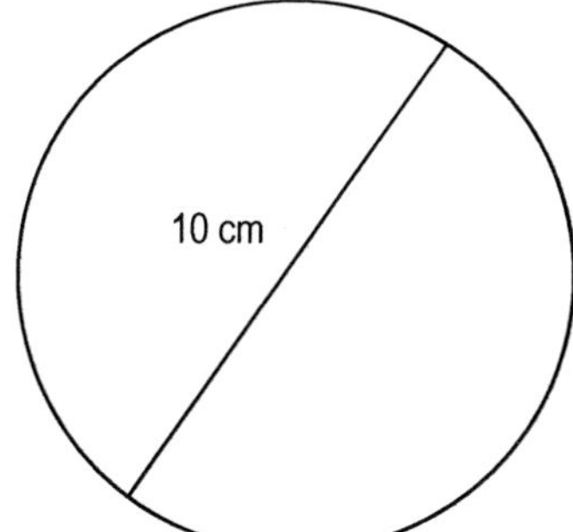

2.

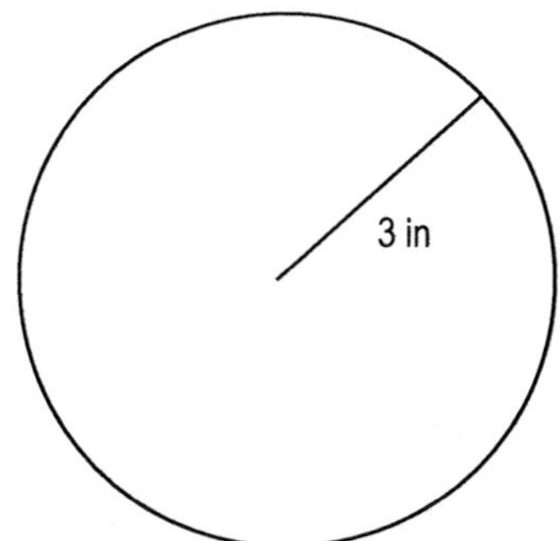

3. Find the area of the following shape.

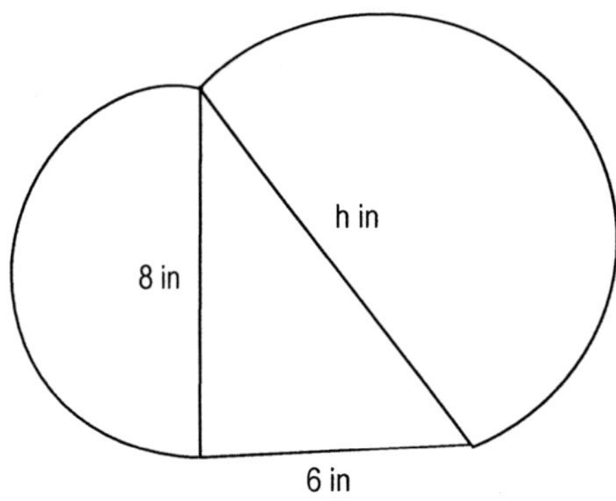

4. Find the area of the following shape.

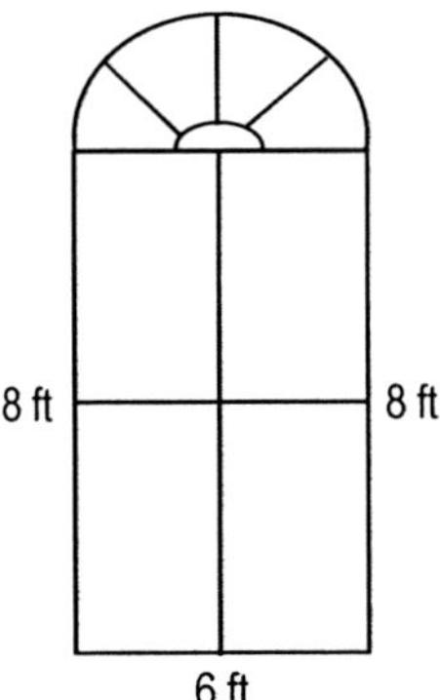

HOMEWORK ASSIGNMENT: 10.6

Name: ______________________________

Find the area and circumference of the following circles:

1.

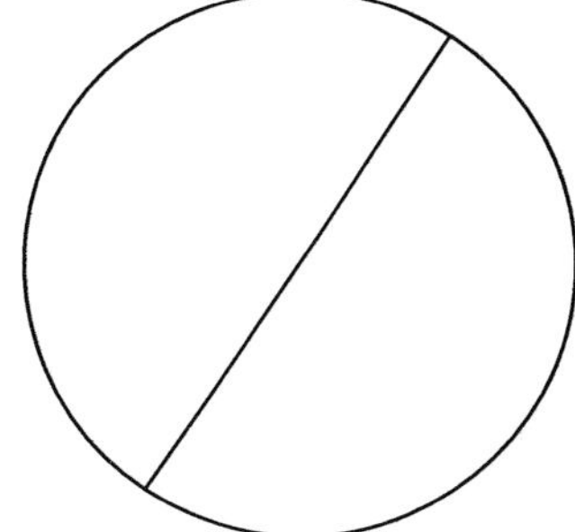

2.

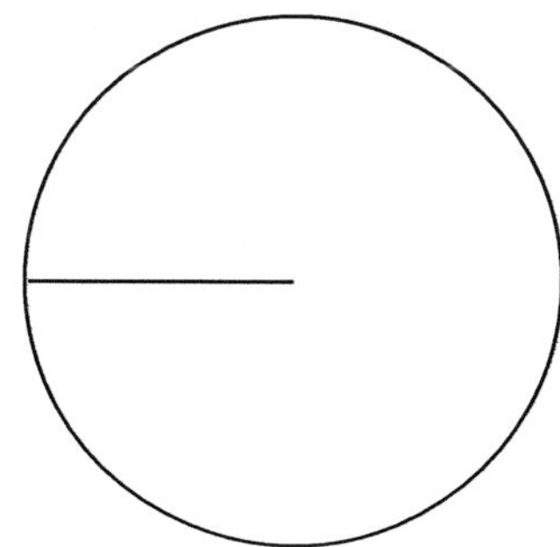

Find the radius given the area or circumference:

3. $A = 379.94 \text{ in}^2$

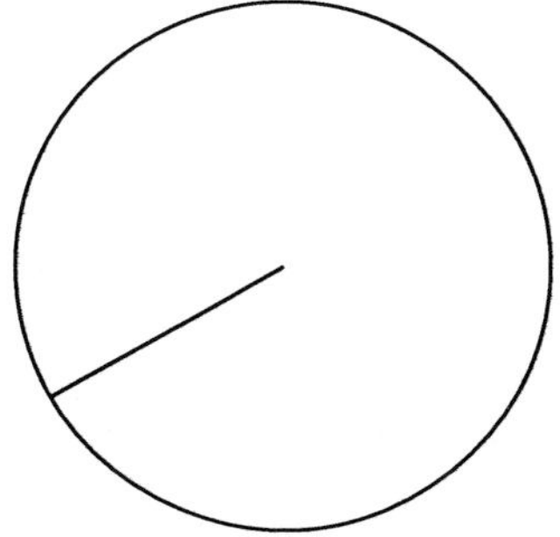

4. $C = 109.9$ yd

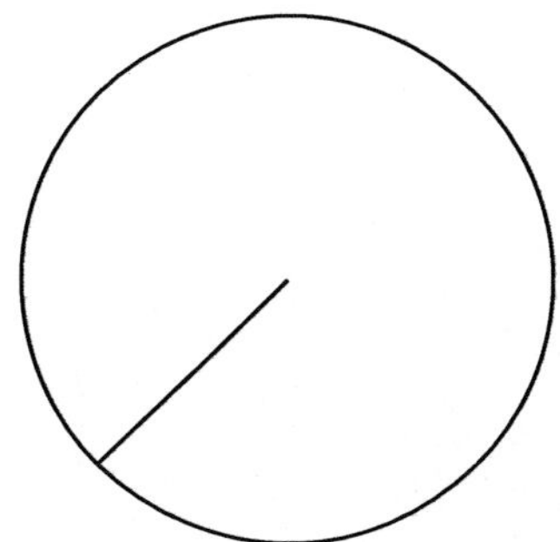

Review:

5. A 24–ounce package of green beans sells for \$1.29. Give the unit cost in cents per ounce.

6. Subtract: $(5x^2 - 8x + 1) - (3x^2 - 2x + 3)$

7. Multiply: $(3x + 2)(2x - 5)$

8. Simplify: $(3a^2 b^4)^3$

10.7 SURFACE AREA AND VOLUME

Volumes and Surface Area of Solids

Figures	Formulas for Volume and Surface Area	
Rectangular Prism	*Volume* = lwh *Surface area* = 2(*length* * *width*) + 2(*length* * *height*) + 2(*width* * *height*) $SA = 2lw + 2lh + 2wh$	
General Prisms	*Volume* = Bh, *where B is the area of the base figure* *Surface area = sum of the areas of all faces*	
Cylinder	*Volume* = πr^2h *Surface area* = $2\pi rh + 2\pi r^2$	$\pi \approx 3.14$ or $\pi \approx$???
Sphere	*Volume* = $\frac{4\pi r^3}{3}$ *Surface area* = $4\pi r^2$	
Right Circular Cone	*Volume* = $\frac{\pi r^2 h}{3}$ *Surface area* = $\pi rl + \pi r^2$, *where l is the slant height*	
Regular Pyramid	*Volume* = $\frac{lwh}{3}$ Surface area = sum of the areas of all faces	

SA of Pyramid with square base: If you break down the shape into smaller basic shapes you find that there are 4 triangles and a square bottom. Use the following formula to keep it simple:

Area of a triangle:

Area of a square:

Use the triangle formula for one triangle and then multiply the result by 4.

Use the square formula for the only base and then add the result to the answer from the triangles calculation.

EXAMPLE:

Find the volume and surface area:

1.

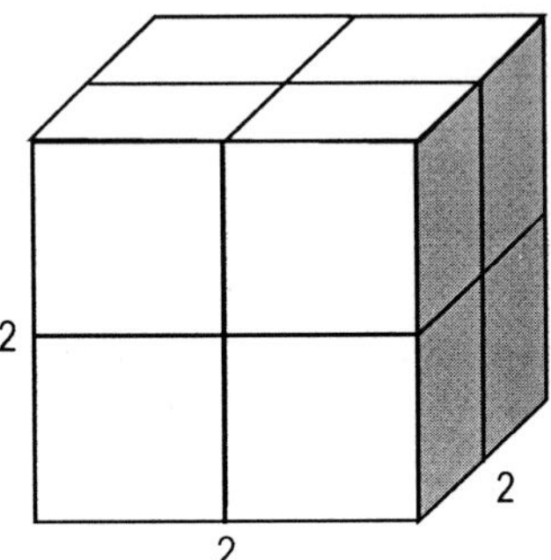

2.

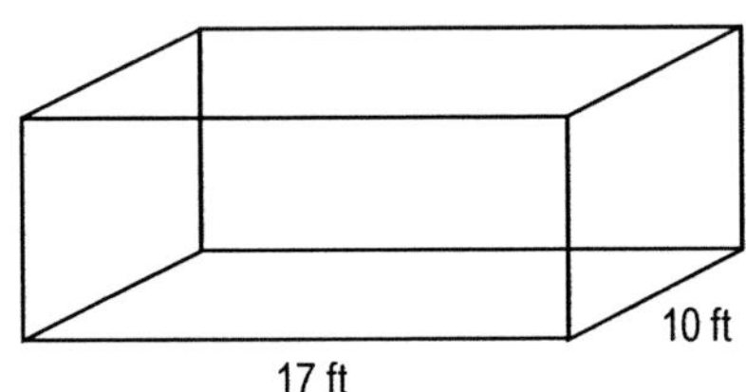

3.

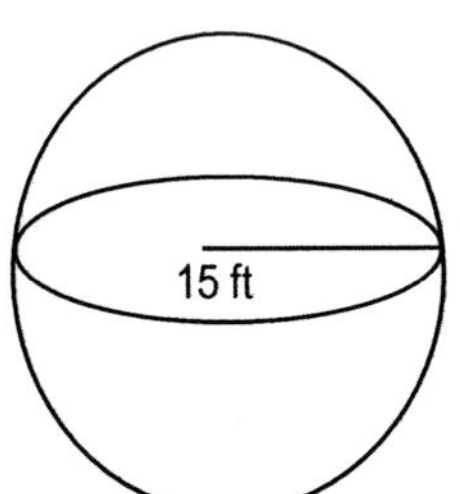

4. Height inside = 9 cm

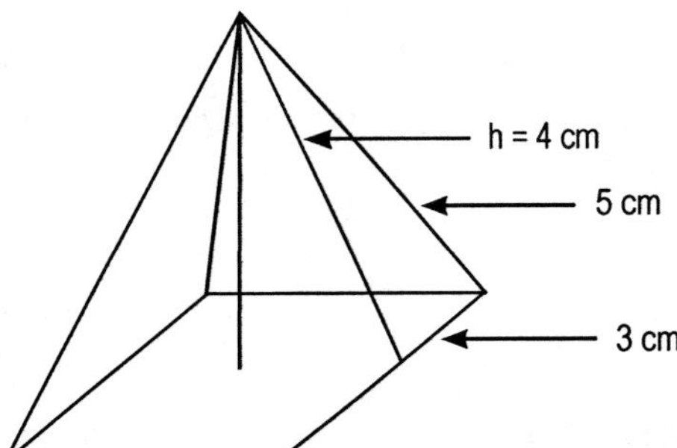

Find the volume:

5.

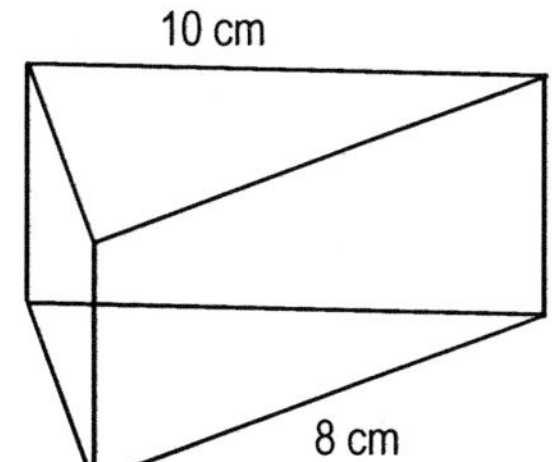

6.

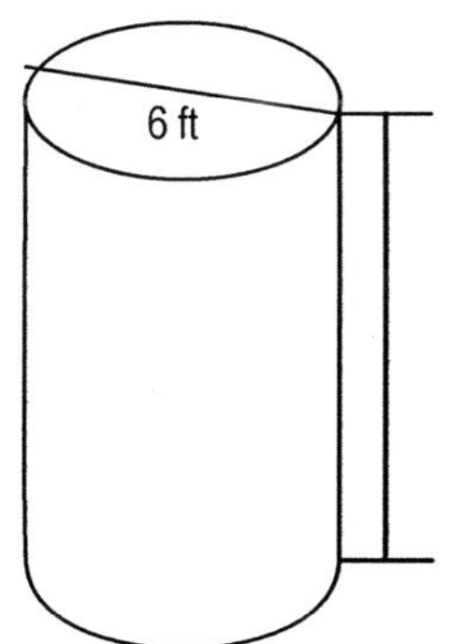

7.

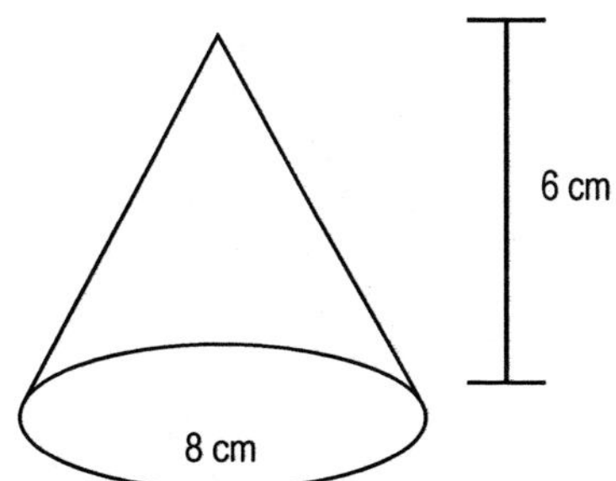

HOMEWORK ASSIGNMENT: 10.7

Name: ______________________________

Find the Volume and Surface Area of $3x4x5$ the following:

1. A rectangular solid with dimensions of .

2. A sphere with a radius of 9 inches.

3. A cylinder with a height of 4 meters and a circular base with a diameter of 18 meters.

4. A cone with a height of 12 centimeters and a circular base with a diameter of 10 centimeters. Find only the volume.

5. A pyramid with a square base of 10 meters on each side and a height of 12 meters.

REFERENCE SHEET

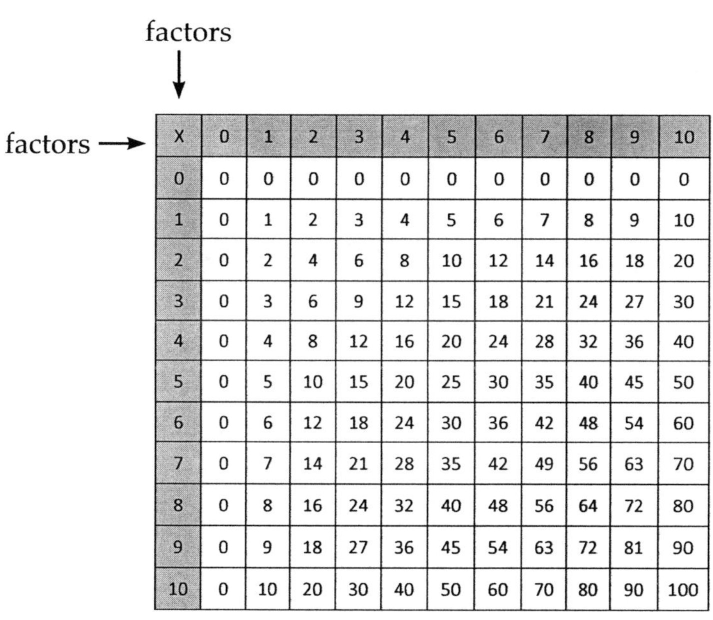

x	0	1	2	3	4	5	6	7	8	9	10
0	0	0	0	0	0	0	0	0	0	0	0
1	0	1	2	3	4	5	6	7	8	9	10
2	0	2	4	6	8	10	12	14	16	18	20
3	0	3	6	9	12	15	18	21	24	27	30
4	0	4	8	12	16	20	24	28	32	36	40
5	0	5	10	15	20	25	30	35	40	45	50
6	0	6	12	18	24	30	36	42	48	54	60
7	0	7	14	21	28	35	42	49	56	63	70
8	0	8	16	24	32	40	48	56	64	72	80
9	0	9	18	27	36	45	54	63	72	81	90
10	0	10	20	30	40	50	60	70	80	90	100

Perfect Squares	Square roots
$1^2 = 1$	$\sqrt{1} = 1$
$2^2 = 4$	$\sqrt{4} = \sqrt{2*2} = 2$
$3^2 = 9$	$\sqrt{9} = \sqrt{3*3} = 3$
$4^2 = 16$	$\sqrt{16} = \sqrt{4*4} = 4$
$5^2 = 25$	$\sqrt{25} = \sqrt{5*5} = 5$
$6^2 = 36$	$\sqrt{36} = \sqrt{6*6} = 6$
$7^2 = 49$	$\sqrt{49} = \sqrt{7*7} = 7$
$8^2 = 64$	$\sqrt{64} = \sqrt{8*8} = 8$
$9^2 = 81$	$\sqrt{81} = \sqrt{9*9} = 9$
$10^2 = 100$	$\sqrt{100} = \sqrt{10*10} = 10$
$11^2 = 121$	$\sqrt{121} = \sqrt{11*11} = 11$
$12^2 = 144$	$\sqrt{144} = \sqrt{12*12} = 12$
$13^2 = 169$	$\sqrt{169} = \sqrt{13*13} = 13$

Simple Interest	$I = prt$
Compound Interest	$A = p\left(1 + \frac{r}{h}\right)^{nt}$

Translations

Addition	Example	Subtraction	Example
1. *sum of p and* 15	$p + 15$	1. *difference of* 30 *an k*	$30 - k$
2. 10 *plus* c	$10 + c$	2. 1,000 *minus* R	$1000 - R$
3. 5 *added to a*	$a + 5$	3. 15 *less than w*	$w - 15$
4. 4 *more than r*	$r + 4$	4. *r decreased by* 5	$r - 5$
5. 8 *greater than A*	$A + 8$	5. *T reduced by* 80	$T - 80$
6. S *increased by* 100	$S + 100$	6. 7 *subtracted from s*	$s - 7$
7. *exceeds L by* 20	$L + 20$	7. 2,000 *less c*	$2000 - c$

Multiplication	Example	Division	Example
1. *product of* 60 *and h*	$60h$	1. *quotient of B and* 5	$\frac{B}{5}$
2. 10 *times A*	$10A$	2. *T divided by* 50	$\frac{T}{50}$
3. *twice w*	$2w$	3. *the ratio of h to* 3	$\frac{n}{8}$
4. $\frac{1}{2}$ of t	$\frac{1}{2}t$	4. *n split into* 8 *equal parts*	$\frac{n}{8}$

Formulas Used in the Real World

Sale price = original – discount	$s = p - d$
Retail price = cost + markup	$r = c + m$
Profit = revenue – cost	$p = r - c$
Distance = rate · time	$d = rt$
Degree Celsius = (⁰F – 32) times five–ninths	$^0\text{C} = \frac{5}{9}(^0\text{F} - 32)$
Degree Fahrenheit = nine–fifths times ⁰C + 32	$^0\text{F} = \frac{9}{5}(^0\text{C} + 32)$
Distance Fallen = 16 (time)^2	$d = 16t^2$
Mean average = sum of values divided by total number of values	$A = \frac{Sun\ of\ values}{\#\ of\ values}$

American Conversions

UNITS LENGTH

To convert from	Use the following	To convert from	Use the following
Feet to inches	$\frac{12\ in.}{1\ ft}$	Inches to feet	$\frac{1\ ft}{12\ in.}$
Yards to feet	$\frac{3\ ft}{3\ yd}$	Feet to yards	$\frac{1\ yd}{3\ ft}$
Yards to inches	$\frac{36\ in}{1\ yd}$	Inches to yards	$\frac{1\ yd}{36\ in.}$
Miles to feet	$\frac{5{,}280\ ft}{1\ mi.}$	Feet to miles	$\frac{1\ mi.}{5{,}280\ ft}$

UNITS OF WEIGHTS

To convert from	Use the following	To convert from	Use the following
Pounds to ounces	$\frac{16\ oz}{1\ lb}$	Ounces to pounds	$\frac{1\ lb}{16\ oz}$
Tons to pounds	$\frac{2{,}000\ lb}{1\ ton}$	Pounds to tons	$\frac{1\ ton}{2{,}000\ lb}$

UNITS OF CAPACITY

To convert from	Use the following	To convert from	Use the following
Cups to ounces	$\frac{8\ fl\ oz}{1\ c}$	Ounces to cups	$\frac{1\ c}{8\ ft\ oz}$
Pints to cups	$\frac{2\ c}{1\ pt}$	Cups to pints	$\frac{1\ pt}{2\ c}$
Quarts to pints	$\frac{2\ pt}{1\ qt}$	Pints to quarts	$\frac{1\ qt}{2\ pt}$
Gallons to quarts	$\frac{4\ qt}{1\ gal}$	Quarts to gallons	$\frac{1\ gal}{4\ qt}$

UNITS OF TIME

To convert from	Use the following	To convert from	Use the following
Minutes to seconds	$\frac{60\ sec}{1\ min}$	Seconds to minutes	$\frac{1\ min}{60\ sec}$
Hours to minutes	$\frac{60\ min}{1\ hr}$	Minutes to hours	$\frac{1\ hr}{60\ min}$
Days to hours	$\frac{24\ hr}{1\ day}$	Hours to days	$\frac{1\ day}{24\ hr}$

Metric Conversions

LENGTH

1 kilometer(km) = 1,000 meters	1 meter = $\frac{1}{1,000}$ kilometer
1 hectometer(hm) = 100 meters	1 meter = $\frac{1}{100}$ hectometer
1 dekameter(dam) = 10 meters	1 meter = $\frac{1}{10}$ dekameter
1 decimeter (dm) = $\frac{1}{10}$ meter	1 meter = 10 decimeters
1 centimeter (cm) = $\frac{1}{100}$ meter	1 meter = 100 centimeters
1 millimeter (mm) = $\frac{1}{1,000}$ meter	1 meter = 1000 millimeters

MASS

1 kilogram (kg) = 1,000 grams	1 gram = $\frac{1}{1,000}$ kilogram
1 hectogram (hg) = 100 grams	1 gram = $\frac{1}{100}$ hectogram
1 dekagram (dag) = 10 grams	1 gram = $\frac{1}{10}$ dekagram
1 decigram (dg) = $\frac{1}{10}$ gram	1 gram = 10 decigrams
1 centrigram (cg) = $\frac{1}{100}$ gram	1 gram = 100 centigrams
1 milligram (mg) = $\frac{1}{1,000}$ gram	1 gram = 1000 milligrams

CAPACITY

1 kiloliter (kL) = 1,000 liters	1 liter = $\frac{1}{1,000}$ kiloliter
1 hectoliter (hL) = 100 liters	1 liter = $\frac{1}{100}$ hectoliter
1 dekaliter (daL) = 10 liters	1 liter = $\frac{1}{10}$ dekaliter
1 deciliter (dL) = $\frac{1}{10}$ liter	1 liter = 10 deciliters
1 centriliter (cL) = $\frac{1}{100}$ liter	1 liter = 100 centiliters
1 milliliter (mL) = $\frac{1}{1,000}$ liter	1 liter = 1000 milliliters
1 millimeter = $1cm^3$ = 1 cc	1000 cc = 1000 mL = 1 liter

Converting between American and Metric Units

Equivalent Lengths	
American to Metric	Metric to American
1 in. ≈ 2.54 cm	1 cm ≈ 0.3937 in.
1 ft ≈ 0.3048 m	1m ≈ 3.2808 ft
1 yd ≈ 0.9144 m	1m ≈ 1.0936 yd
1 mi ≈ 1.6093 km	1 km ≈ 0.6214 mi

Equivalent weights and masses	
American to Metric	Metric to American
1 oz ≈ 28.35 g	1 g ≈ 0.035 oz
1 lb ≈ 0.454 kg	1 kg ≈ 2.2 lb

Equivalent capacities	
American to Metric	**Metric to American**
1 floz ≈ 0.030 L	1 L ≈ 33.8 fl oz
1 pt ≈ 0.473 L	1 L ≈ 2.1 pt
1 qt ≈ 0.946 L	1 L ≈ 1.06 qt
1 gal ≈ 3.785 L	1 L ≈ 0.264 gal

Pythagorean Formula(Rt. Triangles)	$a^2 + b^2 = c^2$

Polygon	Perimeter	Area
Square	$P = 4s$	$A = s^2$
Rectangle	$P = 2l + 2w$	A = bh
Triangle	$P = a + b + c$	$A = \frac{bh}{2}$
Parallelogram	$P = a + b + c + d$	$A = bh$
Trapezoid	$P = a + b + c + d$	$A = \frac{h(b_1 + b_2)}{2}$
Circles	$C = 2\pi r$, or $C = \pi d$	$A = \pi r^2$, $\pi = 3.14$ or $\frac{22}{7}$

Volumes and Surface Area of Solids

Figures	Volume	Surface Area
Rectangular Prism	$V = lwh$	$SA = 2lw + 2lh + 2wh$
General Prisms	$V = Bh$,B is the area of the base figure	SA = sum of the areas of all faces
Cylinder	$V = \pi r2$ h	$SA = 2\pi rh + 2\pi r^2$
Sphere	$V = \frac{4\pi r^3}{3}$	$SA = 4\pi r^2$
Right Circular Cone	$V = \frac{\pi r^2 h}{3}$	$SA = \pi rl + \pi r^2$ where *l*is the slant heigh
Regular Pyramid	$V = \frac{lwh}{3}$	SA = sum of the areas of all faces